SpringerBriefs in Applied Sciences and Technology

Computational Intelligence

Series Editor

Janusz Kacprzyk, Systems Research Institute, Polish Academy of Sciences,
Warsaw, Poland

SpringerBriefs in Computational Intelligence are a series of slim high-quality publications encompassing the entire spectrum of Computational Intelligence. Featuring compact volumes of 50 to 125 pages (approximately 20,000-45,000 words), Briefs are shorter than a conventional book but longer than a journal article. Thus Briefs serve as timely, concise tools for students, researchers, and professionals.

More information about this subseries at http://www.springer.com/series/10618

Rodrick Wallace

Consciousness, Cognition and Crosstalk: The Evolutionary Exaptation of Nonergodic Groupoid Symmetry-Breaking

Rodrick Wallace
Division of Epidemiology
The New York State Psychiatric Institute
Bronx, NY, USA

ISSN 2191-530X   ISSN 2191-5318 (electronic)
SpringerBriefs in Applied Sciences and Technology
ISSN 2625-3704   ISSN 2625-3712 (electronic)
SpringerBriefs in Computational Intelligence
ISBN 978-3-030-87218-2   ISBN 978-3-030-87219-9 (eBook)
https://doi.org/10.1007/978-3-030-87219-9

This Springer imprint is published by the registered company Springer Nature Switzerland AG
The registered company address is: Gewerbestrasse 11, 6330 Cham, Switzerland

Preface

Without invoking panpsychism or identifying consciousness as a weird form of matter, without mind/body dualism, without the *ignis fatuus* of the 'hard problem' and the many other such constructs that haunt contemporary consciousness studies, the asymptotic limit theorems of information and control theories permit the development of mathematical models recognizably similar to the empirical pictures Bernard Baars and others have drawn of high-level mental phenomena. The methodology revolves around constructing an iterated Morse Function free energy analog from information source uncertainties associated with nonergodic sources necessarily 'dual' in a formal sense to cognitive phenomena. This leads to an iterated entropy-analog from which application of the Onsager approximation from nonequilibrium thermodynamics gives large-scale system dynamics, extending earlier results of Wallace (2005) that were restricted to stationary, ergodic systems. We make application to the dynamics of arousal and distraction, and to other examples.

A modified version of the Kadanoff picture of phase transitions in consciousness emerges from the Morse Function itself in a surprisingly standard manner, closely associated with the breaking of groupoid symmetries driven by fundamental equivalence class algebras.

Although this is far indeed from the familiar world of physical theory, it should be possible, on the basis of these probability models, to develop new statistical tools for the analysis of observational and empirical data regarding cognition and consciousness, as presaged by Dretske (1994). Data analysis is, after all, the only possible source of new scientific knowledge as opposed to the new speculation inherent to mathematical models of complex biological, ecological, physiological and social systems and processes (e.g., Pielou 1977).

We find that consciousness in higher animals is a necessarily stripped-down, greatly simplified, high-speed example of much slower, but far richer, biological processes—like immune function and gene expression—that entertain multiple, simultaneous tunable spotlight 'global workspaces' (Wallace 2005, 2012a, b). All such have emerged through evolutionary exaptations of the inevitable crosstalk afflicting information processes through a kind of 'second law' leakage necessarily associated with information as a form of free energy.

This view represents a wrenching reorientation in contemporary consciousness studies, imposing evolutionary theory to debride the subject of various deep, culturally driven, social constructions (e.g., Tononi et al. 2016).

The presentation is not elementary and assumes a certain mathematical sophistication, familiarity with the basic results of control and information theories, and with the current spectrum of research on consciousness, cognition, and their biological underpinnings. Intrepid readers should, however, be able to take this material and run with it.

Bronx, USA   Rodrick Wallace

Acknowledgments The author thanks the Mathematical Consciousness Science Online Seminar for the opportunity to present some of this material, and, most particularly, for perceptive and probing questions.

References

Dretske F (1994) The explanatory role of information. Philos Trans Royal Soc A 349:59–70.

Pielou E (1977) Mathematical Ecology, Wiley, New York

Tononi G, Boly M, Massimini M, Koch C (2016) Integrated information theory: from consciousness to its physical substrate, Nat Rev Neurosci 17:450–461

Wallace R (2005) Consciousness: A Mathematical Treatment of the Global Neuronal Workspace Model, Springer, New York

Wallace R (2012a) Extending Tlusty's rate distortion index theorem method to the glycome: Do even 'low level' biochemical phenomena require sophisticated cognitive paradigms? BioSyst 107:145–152

Wallace R (2012b) Consciousness, crosstalk, and the mereological fallacy: an evolutionary perspective, Phys Life Rev 9:426–453

From the Notebooks of Charles Darwin...

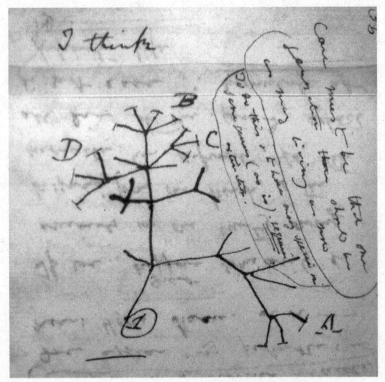

Contents

About the Author

Rodrick Wallace is a research scientist in the Division of Epidemiology at the New York State Psychiatric Institute, affiliated with Columbia University's Department of Psychiatry. He has an undergraduate degree in mathematics and a Ph.D. in physics from Columbia, and completed postdoctoral training in the epidemiology of mental disorders at Rutgers. He worked as a public interest lobbyist and subsequently received an Investigator Award in Health Policy Research from the Robert Wood Johnson Foundation. In addition to material on public health and public policy, he has published peer reviewed studies modeling evolutionary process and heterodox economics, as well as many quantitative analyses of animal, institutional, and machine cognition. He publishes in the military science literature, and in 2019 received one of the UK MoD RUSI Trench Gascoigne Essay Awards.

Chapter 1
Setting the Stage

Nothing in biology makes sense except in light of evolution.
— T. Dobzhansky

Living systems are cognitive systems, and living as a process is a process of cognition.
— Maturana and Varela

1.1 Introduction

The author, as an undergraduate, had a singular conversation with the mathematician John Kemeny, later President of Dartmouth College. Kemeny noted, with some considerable asperity, that physicists were wont to repeatedly publish rediscovered standard results from probability theory, derived using 'childish methods'. The results Kemeny referred to were, of course, the well-known asymptotic limit theorems of the discipline, the Central Limit Theorem, the Renewal Theorem, and so on.

From Dretske (1994) through Tononi et al. (2016), and so on, authors at various levels of scientific sophistication have repeatedly invoked 'communication theory', 'information theory', and similar terms, in various attempts at formal characterization of consciousness. Most often, however, this is done as statement of a—sometimes elaborate—shibboleth rather than as a usable treatment of the subject based on the asymptotic limit theorems of the discipline.

Dretske (1994), by some contrast, provides a—perhaps even the—fundamental insight:

> Unless there is a statistically reliable channel of communication between [a source and a receiver] . . . no signal can carry semantic information . . . [thus] the channel over which the [semantic] signal arrives [must satisfy] the appropriate statistical constraints of communication theory.

© The Author(s), under exclusive license to Springer Nature Switzerland AG 2022
R. Wallace, *Consciousness, Cognition and Crosstalk: The Evolutionary Exaptation of Nonergodic Groupoid Symmetry-Breaking*, SpringerBriefs in Computational Intelligence, https://doi.org/10.1007/978-3-030-87219-9_1

The asymptotic limit theorems of information theory (Cover and Thomas 2006; Khinchin 1957) constrain any and all possible mathematical models of consciousness, and we do well, as Kemeny recognized, to hew closely to them. Embodiment, which is perhaps the most characteristic blindspot afflicting Western consciousness studies, adds another asymptotic limit, the Data Rate Theorem (Nair et al. 2007, described in the Mathematical Appendix), addressing the minimum rate at which control information must be applied to stabilize an inherently unstable system.

Here, we explore ways in which the asymptotic limit theorems of information and control theories can be used to construct models of consciousness that recognizably hew to the picture written by Baars (1989, 2005); Baars and Franklin (2007); Baars et al. (2013) and others as 'global workspace' theory (e.g., Dehaene and Naccache 2001; Dehaene et al. 2011, 2014), models that aid in the creation of new—informed— speculation regarding the phenomenon. The ultimate intent of any such models, however, is the creation of new statistical tools for the analysis of observational and experimental data, the only sources of new knowledge, as opposed to new speculation.

1.2   Some Biological Context

Consciousness is an ancient evolutionary adaptation that provides selective advantage to organisms having identifiable neural systems. It involves a rapid, highly tunable, strongly punctuated, 'spotlight' acquisition to attention and response, typically characterized by a time constant of about 100 ms. Consciousness is not some panpsychic phlogiston, cognitiferous aether, or a new state of matter. The phenomenon gives selective advantage under competition, in spite of the sometimes tenfold metabolic free energy burden that neural structures carry over other tissue forms (Wallace 2012b, and references therein). Consciousness is about an embodied organism interacting with an embedding, and often unfriendly, but highly structured, ecosystem.

That being said, there are, in most organisms, many roughly analogous phenomena, cognitive in the sense of Maturana and Varela (1980), and tunable in very much the same way as consciousness, but acting far more slowly, and—most critically— having multiple simultaneous 'spotlights'. These include

- Glycan code cellular interactions (milliseconds to hours).
- The immune response (hours to days).
- Tumor control (days to years).
- Wound healing (minutes to about 18 months).
- Gene expression (through the life span).
- Institutional and sociocultural group cognition (seconds to centuries).

Earlier work (Wallace 2012b) described these matters as follows,

Cognitive phenomena, it has long been understood... pervade biology, from 'simple' wound healing, through immune maintenance and response, tumor control, neural function, social interaction, and so on. Many can be associated with 'dual' information sources that, following Dretske... are constrained by the necessary conditions imposed by the asymptotic limit theorems of communication theory. Cognition's ubiquity – found at every scale and level of organization of the living state – opens to evolutionary exaptation the crosstalk and noise that plague all information exchange. A principal outcome will be the repeated development of punctuated global broadcast mechanisms that entrain sets of 'unconscious' cognitive modules into shifting, tunable cooperative arrays tasked with meeting the changing patterns of threat and opportunity that challenge all organisms. When such entrainment involves neural systems acting on a timescale of a few hundred milliseconds, the phenomenon is characterized as consciousness. When entrainment involves many individuals or cultural artifacts, the outcomes are social or institutional processes.

Information, however, is also a form of free energy instantiated by physical processes that themselves consume free energy, permitting adaptation of empirical approaches from nonequilibrium thermodynamics and statistical mechanics to cognitive phenomena, given restrictions imposed by the local irreversibility of information sources...

Most singularly, embedding the high level neural global broadcasts of animal consciousness within a nested hierarchy of cognitive and other sources of information evades the logical fallacy of attributing to 'the brain' the broad spectrum of functions that can only be embodied by the full construct of the individual-in-context... The mereological fallacy is, then, fundamentally a matter of decontextualization.

Consciousness has been socially constructed as an 'unsolved scientific mystery'. Another perspective, however, is possible: Since consciousness is constrained to approximately 100 ms response times in neural systems—requiring significant rates of supply of metabolic free energy—higher animals appear to sustain at most a single such 'tunable spotlight'. Slower systems, as they can entertain multiple, interacting, tunable spotlights, are far more complicated (e.g., Wallace 2012a, b). Consciousness is, then, a bare—indeed, stripped-down—version of these richer mechanisms. Such gross simplification is necessary for high-speed response to rapidly shifting patterns of threat and affordance.

1.3   Some Mathematical Context

There are four asymptotic limit theorems of information and control theories that are central to the study of embodied cognition and consciousness:

- The Shannon Coding Theorem (and 'tuning theorem' variants).
- The Source Coding or Shannon-McMillan Theorem.
- The Rate Distortion Theorem (itself a tuning theorem variant).
- The Data Rate Theorem (connecting information and control across inherently unstable systems).

See the Mathematical Appendix for an introduction to the 'tuning theorem' variant of the Coding Theorem, and to the Data Rate Theorem, which are less well known.

These apply to stationary, ergodic phenomena. 'Stationary' means that the probabilities remain constant in time, and 'ergodic' that time averages converge to phase averages. Elsewhere (e.g., Wallace and Wallace 2016), we have expanded consideration to 'adiabatically, piecewise, stationary, ergodic' (APSE) information sources, permitting approximate use of the classic theorems in the same way that the adiabatic approximation in molecular quantum mechanics decouples rapid electron dynamics from much slower nuclear oscillations. Here, however, we will be interested in nonergodic systems, more likely to represent real-world phenomena.

Derivation of analogous—or more precisely, different but relevant—formal results for such systems is no small matter. The Ergodic Decomposition Theorem gives a Ptolemaic system, not a Keplerian or Newtonian one (Hoyrup 2013). The resulting work-arounds generate different classes of Keplerian 'regression model' statistical tools that might be fitted to data.

As Wallace (2018) puts it,

> ...[W]hile every non-ergodic measure has a unique decomposition into ergodic ones, this decomposition is not always computable. From another perspective, such expansions – in terms of the usual ergodic decomposition or the groupoid/directed homotopy equivalents – both explain everything and explain nothing, in the sense that almost any real function can be written as a Fourier series or integral that retains the essential character of the function itself. Sometimes this helps if there are basic underlying periodicities leading to a meaningful spectrum, otherwise not. The analogy is the contrast between the Ptolemaic expansion of planetary orbits in circular components around a fixed Earth vs. the Newtonian/Keplerian gravitational model in terms of ellipses with the Sun at one focus. While the Ptolemaic expansion converges to any required accuracy, it conceals the essential dynamics.

Extending the approach of Wallace (2018), we adapt methods from nonequilibrium thermodynamics to study both nonergodic systems and their ergodic components—when decomposition is exact—in terms of groupoid symmetries associated with equivalence classes of directed homotopy 'meaningful sequences'. The approach is based on recognition of information as a form of free energy rather than as an 'entropy', mathematical form notwithstanding. In the extreme case which will be the starting point individual pathways can be associated with an information source function, but this cannot be represented in terms of a 'Shannon entropy' across a probability distribution.

An equivalence class structure then arises via a metric distance measure—described below in Eq. (4.2)—for which the high probability meaningful sequences of one kind of 'game' are closer together than for a significantly different 'game', Averaging occurs according to such equivalence classes, generating groupoid symmetries. The dynamics are then characterized by symmetry-breaking according to 'temperature' changes that, contrary to Wallace (2018), must be studied from first principles and may incorporate both underlying regulatory mechanisms and the influence of embedding environments. The standard decomposition can, in part, be recovered by noting that larger equivalence classes across which uncertainty measures are constant can be collapsed into single paths on an appropriate quotient manifold.

Finally, we reiterate that, while we can lift the ergodic requirement—that time averages are always global ensemble averages—the theory developed here is very much an 'adiabatic approximation' in the sense that there are local high rate processes that are assumed to 'quasi-equilibrate' while embedding global processes slowly change. That is, we have not fully lifted the 'stationary' requirement surrounding the asymptotic limit theorems of information and control theories (e.g., Khinchin 1957 p. 72), but have attempted to evade it, as stated, in the same way that molecular physics decouples rapid electron dynamics from much slower nuclear vibrations. This remains a central limitation to the work presented here.

1.4 Information and Free Energy

Next, we adapt a page from Feynman (2000, p. 146)—literally—to argue that information can be viewed as a form of free energy. That page recapitulates what Feynman calls a 'quite subtle' argument by Bennett, showing how to construct an ideal engine to convert the information within a message into work, provided the system is 'ergodic', that is, equating time averages to phase averages. Bennett's machine is displayed in Fig. 1.1.

The 'message tape' is fed into the wheeled engine isothermally, generating an average force against the piston, allowing the extraction of work.

There is a more direct way to see this, following the pattern of Wilson (1971). For a physical system of volume V and partition function $\mathscr{Z}(K_1, \ldots, K_m, V)$, where the K_j are parameters and V is the volume, the free energy can be defined as

$$F(K_1, \ldots, K_m) = \lim_{V \to \infty} \frac{\log[\mathscr{Z}(K_1, \ldots, K_m, V)]}{V} \qquad (1.1)$$

For a stationary, ergodic information source, according to the Shannon-McMillan Source Coding Theorem, system paths—messages, in a large sense—can be divided into two sets, one of high probability consonant with underlying grammar and syntax, and one of vanishingly small probability not so consonant (Khinchin 1957).

Let $N(n)$ be the number of high probability grammatical/syntactic paths of length n. Then the Shannon uncertainty of the information source X can be written as

$$H[X] = \lim_{n \to \infty} \frac{\log[N(n)]}{n} \qquad (1.2)$$

For stationary, ergodic information sources dual to cognitive processes one can construct a dynamic theory based simply on this homology (Wallace 2005, 2012b):

- Define an entropy as the Legendre transform $S \equiv -H(\mathbf{K}) + \mathbf{K} \cdot \nabla_{\mathbf{K}} H$ and impose the dynamics of first-order Onsager nonequilibrium thermodynamics, understanding that there is no microreversibility for information transmission, and hence no 'Onsager reciprocal relations'.

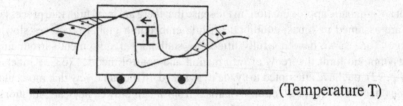

(Temperature T)

An Information-driven Engine

We now let the heat bath warm the cell up. This will cause the atom in the cell
to jiggle against the piston, isothermally pushing it outwards.

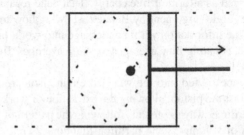

Work Generation Mechanism in the Engine

Fig. 1.1 Adapted from Feynman (2000). Bennett's ideal machine converting information within a
message to work—free energy

- Impose Wilson's (1971) version of the Kadanoff model on H, using 'biological
  renormalizations' (Wallace 2005).

We are going to be interested here, however, in nonergodic systems more likely
to characterize the real world, that is, systems for which time averages are not given
by phase averages. This requires some serious thought.

1.5 Basic Variables

An embodied cognitive agent is embedded in, acting on, and acted on by, a landscape
of imprecision in effect, reaction, and result.

The agent enjoys three essential resource streams. These are, first, the rate at
which information can be transmitted between parts of itself, characterized by an
information channel capacity $\mathscr{C}$. The second resource stream is the rate at which
sensory information about the embedding environment is available, at some rate $\mathscr{H}$.

The third is the rate at which metabolic free energy and related 'real' resources can be delivered, $\mathcal{M}$.

The resource rates, and time, will interact, generating a correlation matrix analog $\mathbf{Z}$ of dimension 3. Any n dimensional matrix has n scalar invariants—characteristic numbers that remain the same under certain transformations. These invariants can be found from the standard polynomial relation

$$p(\gamma) = \det[\mathbf{Z} - \gamma \mathbf{I}] = \gamma^n - r_1 \gamma^{n-1} + r_2 \gamma^{n-2} - \cdots + (-1)^n r_n \qquad (1.3)$$

Here, $\mathbf{I}$ is the n-dimensional identity matrix, det the determinant, and γ a real-valued parameter. The first invariant is usually taken as the matrix trace, and the last as $\pm$ the matrix determinant.

These invariants can be used to build a single scalar index $Z = Z(r_1, \ldots, r_n)$. The simplest such would be $Z = \mathcal{C} \times \mathcal{H} \times \mathcal{M}$. Scalarization, however, must be appropriate to the system under study at the time of study, and there will almost always be cross-interactions between these rates.

The important point for this analysis is that scalarization permits analysis of a one dimensional system. Expansion of Z into vector form leads to sometimes difficult multidimensional dynamic equations (Wallace 2020a, Sect. 7.1). See the Mathematical Appendix for an outline.

1.6 On Integrated Information Theory (IIT), or Down the Rabbit Hole

Finally, a trick—but hopefully instructive—question: In the spirit of John Kemeny, what, and under what conditions, is the asymptotic limit theorem satisfied by the 'integrated information' measure of Tononi et al. (2016)?

As Hanson and Walker (2021) put it,

...[IIT]... demonstrates several hallmark features of an unscientific handling of ideas... The first is the use of an ad hoc procedure to resolve fundamental contradictions inherent in the theory... The second is a lack of transparency with regard to the mathematical implementation of the theory... And third, the theory struggles to ground itself experimentally... In combination, these three features strongly suggest that IIT is on the wrong side of the demarcation problem... [Resolving such matters]... is especially important given the negative repercussions that could result from the misapplication of a theory of consciousness in medical, legal, and moral settings.

These are not trivial complaints, and similar misgivings regarding IIT have been raised by others (e.g., Moon 2019). The question, then, remains: even if the 'phi' of IIT could be calculated reliably, what asymptotic limit theorem does it satisfy? This, too, is not a trivial matter. Strategies of scientific inquiry are confronted by the same recurring theme that confounds strategy in most other human enterprise: 'So What!' (Gray 2018). Even if some more rigorously-defined version of Φ can be calculated, So What!

Doerig et al. (2019) encapsulate criticisms of IIT in terms of an 'unfolding argument' that does not apply to Baars-type approaches to consciousness:

Certain theories suggest that consciousness should be explained in terms of brain functions, such as accessing information in a global workspace, applying higher order to lower order representations, or predictive coding. These functions could be realized by a variety of patterns of brain connectivity. Other theories, such as Information Integration Theory (IIT) and Recurrent Processing Theory (RPT), identify causal structure with consciousness. For example, according to these theories, feedforward systems are never conscious, and feedback systems always are. ...[U]sing theorems from the theory of computation, we show that causal structure theories are either false or outside the realm of science [since 'conscious' systems can always be replaced with functionally-identical 'nonconscious' structures].

Herzog et al. (2021) assert that, since IIT's level and content of consciousness are fully dissociated from any behavioral i/o function, IIT's 'consciousness' is acausal, without any impact on the world.

Tegmark (2016), in a de-facto, if inadvertent, *reducto ad absurdum*, catalogs some hundreds of possible Φ-like information integration measures, finding six that 'stand out' as having reasonable properties and being, for the most part, actually subject to calculation. Again, one can ask, So What! That is, (1) what are the asymptotic limit theorems that these measures satisfy, and (2) are any such theorems actually of use in constructing statistical tools for data analysis?

Here, we will focus precisely on 'what it does', revolving around the evolutionary exaptation of information crosstalk across a variety of biological venues, leading to probability models associated with asymptotic limit theorems that can be developed into statistical tools for the analysis of empirical and observational data. Elsewhere, we have used similar methods to explore the dynamics of machine intelligence and institutional cognition (e.g., Wallace 2020a, b).

References

Baars B (1989) A cognitive theory of consciousness. Cambridge University Press, New York
Baars B (2005) Global workspace theory of consciousness: toward a cognitive neuroscience of human experience. Prog Brain Res 150:45–53
Baars B, Franklin S (2007) An architectural model of conscious and unconscious brain functions: global workspace theory and IDA. Neural Netw 20:955–961
Baars B, Franklin S, Ramsoy T (2013) Global workspace dynamics: cortical binding and propagation enables conscious contents. Front Psychol 4:200
Cover T, Thomas J (2006) Elements of information theory, 2nd edn. Wiley, New York
Dehaene S, Naccache L (2001) Towards a cognitive neuroscience of consciousness: basic evidence and a workspace framework. Cognition 79
Dehaene S, Changeux JP, Naccache L (2011) The global neuronal workspace model of conscious access: from neuronal architectures to clinical applications. In: Dehaene S, Christen Y (eds) Characterizing consciousness: from cognition to the clinic?. Springer, Berlin, pp 55–84
Dehaene S, Charles L, King J, Marti S (2014) Toward a computational theory of conscious processing. Curr Opin Neurobiol 25:76–84
Doerig A, Schurger A, Hess KI, Herzog M (2019) The unfolding argument: why IIT and other causal structure theories cannot explain consciousness. Conscious Cogn 72:49–59

Dretske F (1994) The explanatory role of information. Philos Trans R Soc A 349:59–70

Feynman R (2000) Lectures on computation. Westview Press, New York

Gray C (2018) So what! the meaning of strategy. Infin J 6(1):4–7. https://www.militarystrategymagazine.com/article/so-what-the-meaning-of-strategy/

Hanson J, Walker S (2021) On the non-uniqueness problem in integrated information theory. https://doi.org/10.1101/2021.04.07.438793v1

Herzog M, Schurger A, Doerig A (2021) Pure first-person experience and the unfolding argument: Neo-Cartesian reasoning as a foundation for IIT and other causal structure theories ends up in dissociative epiphenomenalism. https://psyarxiv.com/s8a7n/

Hoyrup M (2013) Computability of the ergodic decomposition. Ann Pure Appl Log 164:542–549

Khinchin A (1957) Mathematical foundations of information theory. Dover, New York

Maturana H, Varela F (1980) Autopoiesis and cognition: the realization of the living. D. Reidel Publishing, Dordrecth

Moon K (2019) Exclusion and underdetermined qualia. Entropy 21:405

Nair G, Fagnani F, Zampieri S, Evans R (2007) Feedback control under data rate constraints: an overview. Proc IEEE 95:108–137

Tegmark M (2016) Improved measures of integrated information. PLOS Comput Biol. https://doi.org/10.1371/journal.pcbi.1005123

Tononi G, Boly M, Massimini M, Koch C (2016) Integrated information theory: from consciousness to its physical substrate. Nat Rev Neurosci 17:450–461

Wallace R (2005) Consciousness: a mathematical treatment of the global neuronal workspace model. Springer, New York

Wallace R (2012a) Extending Tlusty's rate distortion index theorem method to the glycome: do even 'low level' biochemical phenomena require sophisticated cognitive paradigms? BioSystems 107:145–152

Wallace R (2012b) Consciousness, crosstalk, and the mereological fallacy: an evolutionary perspective. Phys Life Rev 9:426–453

Wallace R (2018) New statistical models of nonergodic cognitive systems and their pathologies. J Theor Biol 436:72–78

Wallace R (2020a) How AI founders on adversarial landscapes of fog and friction. J Def Model Simul. https://doi.org/10.1177/1548512920962227

Wallace R (2020b) Cognitive dynamics on Clausewitz landscapes: the control and directed evolution of organized conflict. Springer, New York

Wallace R, Wallace D (2016) Gene expression and its discontents: the social production of chronic disease, 2nd edn. Springer, New York

Wilson K (1971) Renormalization group and critical phenomena. I renormalization group and the Kadanoff scaling picture. Phys Rev B 4:317–83

Chapter 2
Embodied Cognition and Its Dynamics

2.1 The First Round

Embodied entities are built from crosstalking cognitive submodules. These range from individual cells, organs, social groupings, formal institutions, to embedding cultures and other environments. For humans in particular, every scale and level of organization, individuals and their social workgroups are constrained, not only by their own experience and training, but by the culture in which they are embedded and with which they interact (e.g., Wallace 2018, 2020).

They are likewise constrained by the the environment in which they operate, including actions and intents of competing and cooperating entities.

Further, there is always structured uncertainty imposed by the large deviations possible within the overall system, including, but not limited to, the embedding environment.

Thus, a number of factors interact to build a composite information source (Cover and Thomas 2006) representing embodied cognition. These are

- Cognition requires choice that reduces uncertainty and implies the existence of an information source formally 'dual' to that cognition at each scale and level of organization (Atlan and Cohen 1998). The argument is direct and agnostic about representation.
- Cognition requires regulation. As Wallace (2017, Chap. 3) puts it,

  Cognition and its regulation ... must be viewed as an interacting gestalt, involving not just an atomized individual, but the individual in a rich context... There can be no cognition without regulation, just as there can be no heartbeat without control of blood pressure, and no multicellularity without control of rogue cell cancers. Cognitive streams must be constrained within regulatory riverbanks.

© The Author(s), under exclusive license to Springer Nature Switzerland AG 2022
R. Wallace, *Consciousness, Cognition and Crosstalk: The Evolutionary Exaptation of Nonergodic Groupoid Symmetry-Breaking*, SpringerBriefs in Computational Intelligence, https://doi.org/10.1007/978-3-030-87219-9_2

It is here that the Data Rate Theorem, or an appropriate generalization, becomes manifest: there must be an embedding regulatory information source imposing control information at a rate greater than an inherently unstable cognitive process generates its own 'topological information'. See the Mathematical Appendix for details.

- For humans in particular, embedding culture is also an information source, with analogs to grammar and syntax: within a culture, under particular circumstances, some sequences of behavior are highly probable, and others have vanishingly small probability (Khinchin 1957), a sufficient condition for the development of an equivalence class groupoid symmetry-breaking formalism.
- Spatial and social geographies are similarly structured so as to have incident sequences of very high and very low probability: night follows day, summer's dirt roads are followed by October's impassible mud streams.
- Large deviations, as described by Champagnat et al. (2006) and Dembo and Zeitouni (1998), follow high probability developmental pathways governed by entropy-like laws that imply the existence of another information source.

Embedded and embodied cognition is then characterized by a joint information source uncertainty (Cover and Thomas 2006) as

$$H(\{X_i\}, X_V, X_\Delta) \tag{2.1}$$

The set $\{X_i\}$ includes the cognitive, regulatory, and embedding cultural information sources of the hierarchical system, X_V is the information source of the embedding environment, that may include the actions and intents of adversaries or collaborators, as well as 'weather'. Finally, X_Δ is the information source of the associated large deviations possible to the system.

The essential point is that crosstalk between coresident information channels and sources is almost inevitable, a consequence of the information chain rule (Cover and Thomas 2006). For a set of interacting stationary ergodic information sources X_i, $i = 1, 2, \ldots$, the joint uncertainty of interacting sources and channels is always less than or equal to the sum of the independent uncertainties (Cover and Thomas 2006):

$$H(X_1, X_2, \ldots) \leq \sum_i H(X_i) \tag{2.2}$$

Each information source X_i is powered by some corresponding free energy source $\mathcal{M}_i$, and it takes more free energy to isolate information sources and channels than to allow their interaction. Such a 'second law' conundrum confounds much of electrical engineering, particularly in the design and construction of microchips. Such 'second law' problems extend to all scales and levels of organization.

Evolutionary process has taken this 'spandrel', in the sense of (Gould and Lewontin 1979), and built whole new cathedrals from it (Wallace 2012).

The next steps are somewhat subtle.

According to popular mathematical canon, there is really no serious work to be done on nonergodic information sources as a consequence of the Ergodic Decomposition Theorem (e.g., Gray 1988 Chap. 7) which states that it is possible to factor any nonergodic process into a sufficiently large sum (or generalized integral) of ergodic processes, in the same way that any point on a triangle can be expressed in terms of its extremal fixed point vertexes. As Winkelbauer (1970) put it for information source uncertainty,

Theorem II. *The asymptotic rate of a stationary source* μ *equals the essential supremum of the entropy rates of its ergodic components:*

$$H(\mu) = ess. \sup_{z \in R[\mu]} H(\mu_z)$$

where the μ_z are ergodic.

Is this really a 'simple' result for dynamic systems that can suffer 'absorbing states'? Individual paths—and small, closely-related, equivalence classes of them—are particularly important in biological phenomena, as opposed to physical process. This is because each path may have a unique consequence for the organism or other entity embedded in a stressful environment. After all, there will only be a single 'meaningful sequence' associated with successful capture by a predator. That is, absorbing states are particularly important in biological processes.

Recall, further, that it is possible to approximate any reasonably well-behaved real-valued function over a fixed interval in terms of a Fourier series. Recall that it was, in the geocentric Ptolemaic system, via a sufficient number of epicycles, possible to predict planetary positions to any desired accuracy using such a de-facto Fourier Decomposition. The underlying astronomical problem was both considerably simplified and greatly enhanced by the non-geocentric empirical observations of Kepler, explained by Newton, and fully elaborated by Einstein.

The phenomena of cognition and consciousness are considerably more complex than the motion of the planets around the sun, and Keplerian laws must still be found across many different physiological phenomena and organisms. Newton and Einstein are nowhere on the horizon for theories of cognition and consciousness.

Here, we significantly expand the development of Wallace (2018), deriving from first principles, rather than imposing, a 'temperature' measure for nonergodic cognitive systems.

We have, above, reduced the spectrum of resources and their interactions—including internal bandwidth, rates of sensory information, and material/energy supply—in terms of a scalar rate variable Z.

To explore some dynamic processes, we next introduce a first-order linear Onsager approximation abducted from nonequilibrium thermodynamics (de Groot and Mazur 1984).

Here, we invoke an *iterated* free energy Morse Function (Pettini 2007) via a formalized Boltzmann probability expression, in the sense of Feynman (2000). This is done by enumerating high probability developmental pathways available to the system as $j = 1, 2, \ldots$, allowing definition of a path probability P_j

$$P_j = \frac{\exp[-H_j/g(Z)]}{\sum_k \exp[-H_k/g(Z)]} \tag{2.3}$$

This formulation, following Khinchin (1957) and Wallace (2018), will apply to stationary nonergodic as well as to stationary ergodic information sources and can be used for systems in which each developmental pathway x_j has its own source uncertainty measure H_{x_j}. This, however, is only expressed as a Shannon 'entropy' for an ergodic system (Khinchin 1957).

We are, in some measure, liberated from the kind of dilemma Tegmark (2016) explores for Integrated Information Theory in that we do not need the functional form of a H_k, only its scalar magnitude. Further development, however, involves imposition of an 'adiabatic approximation' that assumes a 'sufficiently stationary' convergence about slowly-varying underlying system dynamics, much as is done in the study of molecular quantum mechanics, where rapidly-changing electron dynamics are approximated around slowly-changing nuclear configurations. Previous work—Wallace (2005, 2012)—specified 'adiabatically, piecewise, stationary ergodic' (APSE) systems that remained as close to stationary and ergodic as needed for the asymptotic limit theorems to work. Here, we lift the ergodic requirement.

The 'temperature' $g(Z)$, like the loss of ergodicity, also becomes another matter, and must now be calculated from Onsager-like system dynamics built from the partition function, i.e., from the denominator of Eq. (2.3). This implies its own set of approximations.

The system's 'rate of cognition' can then be expressed, as in chemical reaction theory (Laidler 1987), by the probability such that $H_j > H_0$, where H_0 is the lower limit for detection of a signal under embodiment in a varying and noisy environment, or for stability via the Data Rate Theorem, as described in the Mathematical Appendix.

There is an important point implicit here: The 'prime groupoid phase transition'. We are, in essence, imposing a symmetry-breaking version of the Source Coding Theorem, the Shannon-McMillan Theorem, on the system, dividing all possible paths into a small set—an equivalence class—of 'meaningful' sequences consonant with some underlying grammar and syntax—in a large sense—and a very large set of vanishingly small probability paths that are not consonant.

Such an equivalence class partition imposes the first of many groupoid symmetry breaking phase transitions. We will discuss groupoid symmetries in more detail below, but, in essence, we have used a groupoid symmetry-breaking phase change to derive—or to impose—a fundamental information theory asymptotic limit theorem. Other such may emerge 'naturally' from related groupoid symmetry-breaking phenomena, all related to equivalence classes of system developmental pathways.

This result can be seen, from a biological perspective, as in the same ballpark as a much earlier phase transition, the sudden transmission of light across the primordial universe after the first 370,000 years.

In sum, groupoid symmetry-breaking extends the Shannon-McMillan Source Coding Theorem to nonergodic—and possibly non-stationary—information sources. This is a major—if 'trivially obvious'—result to which we will return below. The Mathematical Appendix provides a brief introduction to the standard groupoid algebra (Brown 1992; Cayron 2006; Weinstein 1996). The central matter is that products are not defined for all possible pairs of elements, leading to disjoint orbit partition.

The iterated free energy Morse Function F is defined by the relation

$$\exp[-F/g(Z)] \equiv \sum_k \exp[-H_k/g(Z)] = h(g(Z)) \qquad (2.4)$$

F is a Morse Function subject to symmetry-breaking transitions as $g(Z)$ varies (Pettini 2007; Matsumoto 1997). See the Mathematical Appendix for an outline of Morse Function formalism.

We reiterate that these symmetries are not those associated with simple physical phase transitions represented by standard group structures. Cognitive phase change involves punctuated transitions between equivalence classes of high probability signal sequences, represented as groupoids. As Tateishi et al. (2013) put it, if experimental data can be grouped into equivalence classes compatible with an algebraic structure, a groupoid approach can capture the symmetries of the system in a way not be possible with group theory, for example in the analysis of neural network dynamics. Deeper delvings into similar matters can be found in Schreiber and Skoda (2010).

In this work, groupoid symmetries are driven by the directed homotopy induced by failure of local time reversibility for information systems. This is because palindromes have vanishingly small probability. In English, 'the' has meaning in context while 'eht' has vanishingly low probability. The ubiquitous mathematical trope 'except on a set of measure zero' implies a fundamental symmetry breaking.

Particularly complicated cognitive systems may require even more general structures, for example, small categories and/or semigroupoids for analogs to the standard symmetry-breaking dynamics of physical systems.

These more general symmetry-breaking phase changes represent extension of the Data Rate Theorem (DRT) to cognitive systems. Again, the DRT states that the rate at which externally-supplied control information must be imposed on an inherently unstable system to stabilize it must exceed the rate at which that system generates its own 'topological information'. The model is of steering a vehicle on a rough, twisting roadway at night. The headlight/steering/driver complex must impose control information at a rate greater than the 'twistiness/roughness' of the road imposes its own information on the vehicle.

There may, then, be many phase analogs available to a cognitive system as $g(Z)$ varies, rather than just the 'on/off' of stability implied by the DRT itself. We will make something of this in a following section that generalizes 'renormalization' analysis of phase transition.

Dynamic equations can be derived from from Eq. (2.3) by invoking a first order Onsager approximation in the gradient of an entropy measure constructed from the iterated free energy Morse Function F via the Legendre transform

$$S(Z) \equiv -F(Z) + ZdF/dZ \qquad (2.5)$$

After some development,

$$\exp[-F/g(Z)] =$$
$$\sum_k \exp[-H_k/g(Z)] \equiv h(g(Z))$$
$$F(Z) = -\log(h(g(Z))g(Z)$$
$$g(Z) = -\frac{F(Z)}{RootOf\left(e^X - h(-F(Z)/X)\right)}$$
$$\partial Z/\partial t \approx \mu \partial S/\partial Z = f(Z) \qquad (2.6)$$

where the *RootOf* construct defines a generalized Lambert W-function in the sense of Maignan and Scott (2016), Mezo and Keády (2015), and Scott et al. (2006).

The last expression in Eq. (2.6) represents imposition of the entropy gradient formalism of Onsager nonequilibrium thermodynamics (de Groot and Mazur 1984), for which $f(Z)$ is the 'adaptation function', the fundamental rate at which the system adjusts to changes in Z. Again, for information transmission there is no 'temporal microreversibility', so that there can be no Onsager Reciprocal Relations in these models.

After some further work,

$$f(Z) = Zd^2F/dZ^2$$
$$F(Z) = \int\int \frac{f(Z)}{Z}dZdZ + C_1Z + C_2$$
$$-Z\int \frac{f(Z)}{Z}dZ - \log(h(g(Z)))g(Z) - C_1Z + \int (f(Z)dZ + C_2 = 0 \quad (2.7)$$

Again, taking $F = -\log(h(g(Z))g(Z)$, with h determined by underlying internal structure, leads to expressing g in terms of a generalized Lambert W-function, suggesting an underlying formal network structure (Newman 2010).

Specification of any two of f, g, h, in theory, allows calculation of the third. Note, however, that h is fixed by the internal structure of the larger system, and the 'adaptation rate' f is imposed by externalities. In addition, the 'boundary conditions' C_1, C_2 are likewise externally-imposed, also structuring the temperature-analog $g(Z)$. Indeed, the 'temperature' $g(Z)$ might well be viewed as itself an order parameter.

We assume here that embodied cognitive systems can be characterized by the scalar parameter Z, mixing material resource/energy supply with internal and external flows of information under time constraint. There may be more than one such composite irreducible entity driving system dynamics. More explicitly, it may be necessary to replace the scalar Z with some $m \leq n$-dimensional vector having a number of independent—even orthogonal—components accounting for considerable portions of the total variance in the rate of supply of essential resources. The dynamic equations can then be reexpressed in a more complicated vector form. See the Mathematical Appendix for an outline.

In a similar way, it may be necessary to introduce nonlinear or higher order Onsager models. An introduction to these matters can be found in Wallace (2021), involving, for example, expressions of the form

$$S = -F + \sum_j a^j Z^j d^j F/dZ^j$$

$$\partial Z/\partial t \approx \sum_j b_j d^j S/dZ^j = f(Z)$$

$$\partial Z/\partial t \approx dS/dZ \times d^2 S/dZ^2 + \cdots = f(Z) \tag{2.8}$$

leading to formal algebraic power series treatments in the sense of Jackson et al. (2017) that we will explore in more detail in the next chapter. These considerations take us beyond the simplest '$Y = mX + b$' regression model analogs.

The dynamics are driven at rates determined by the adaptation function $f(Z)$. We can ask more detailed questions regarding what happens at critical points defined in terms of the 'temperature' variate $g(Z)$ through the abduction of another approach from physical theory.

2.2 The Second Round

We have, for cultural and historical reasons, focused here largely on rapid, single-workspace phenomena of neural consciousness having the 100 ms time constant. If we enthrone the Data Rate Theorem rather than rate of cognition, we can say something regarding embodied cognition as a (or even the) more fundamental gestalt. The argument—a significant condensation of Wallace (2021)—is surprisingly simple, and seems independent of such niceties as ergodicity of information sources.

Recall the Data Rate Theorem—e.g., Fig. 6.1 in the Mathematical Appendix. Stabilization of an inherently unstable control system engaged in some fundamental task requires that control information be delivered at a rate greater than the rate at which the unstable system generates its own 'topological information', say H_0. We do not specify 'H_0', except as a scalar entity that may indeed change with time.

Most simply, we characterize H as the (scalar) rate at which external regulatory mechanisms provide such control information, and (yet again) define a Boltzmann pseudoprobability

$$dP(H) \equiv \frac{\exp[-H/g(Z)]dH}{\int_0^\infty \exp[-H/g(Z)]dH} \tag{2.9}$$

where $g(Z)$ and Z are as described above, supposing that the regulatory-action process is itself composed of subcomponents that interact with each other and with an embedding environment through both information exchanges and the use of 'materiel' in various forms.

We define iterated free energy and entropy analogs as above from the 'partition function' denominator of Eq. (2.9), obtaining the usual relations

$$\exp[-F/g(Z)] \equiv \int_0^\infty \exp[-H/g(Z)]dH = g(Z)$$

$$g(Z) = \frac{-F(Z)}{W(n, -F(Z))}$$

$$S \equiv -F + ZdF/dZ$$

$$\partial Z/\partial t \propto dS/dZ = f(Z)$$

$$f(Z) = Zd^2F/dZ^2$$

$$L(Z) = \frac{\int_{H_0}^\infty \exp[-H/g(Z)]dH}{\int_0^\infty \exp[-H/g(Z)]dH} = \exp[-H_0/g(Z)] \tag{2.10}$$

where, again, $W(n, x)$ is the appropriate Lambert W-function and $L(Z)$ the cognition rate. A more convoluted line of argument leads to the 'generalized Lambert W-functions' of Eq. (2.6).

Embodiment—directly implying interaction with the embedding world—appears to impose a certain draconian or Procrustean simplicity, as Charles Darwin, Alfred Russel Wallace, and many others have noted.

2.3   Toward a Third Round

Another possible approach is via a Data Rate Theorem (DRT) implementation of the Rate Distortion Theorem (RDT) for non-ergodic information sources. The DRT is briefly discussed in the Mathematical Appendix, and we refer to Fig. 6.1, following how a control signal u_i is expressed in the system response at time $t + 1$ as x_{t+1}. Adopting the RDT perspective, we deterministically retranslate an observed set of outputs $X^i = \{x_1^i, x_2^i, \ldots\}$ into a sequence of possible control signals $\hat{U}^i = \{\hat{u}_1^i, \hat{u}_2^i, \ldots, \}$ and compare the inferred control sequence with the original control sequence $U^i = \{u_1^i, u_2^i, \ldots\}$. A 'difference' between the U is scalarized under the 'distortion measure' H, the control information rate necessary to impose stability on the inherently unstable control system. We then define an 'average distortion' $\mathscr{H}$ as

$$\mathscr{H} = \sum_i p(U^i)H(U^i, \hat{U}^i) \tag{2.11}$$

where $p(U^i)$ is the probability of the sequence u^i. $H(U^i, \hat{U}^i)$, the external control information rate needed for stability, is to be taken as the 'information distortion' between the two control sequences.

It is possible to define a convex Rate Distortion Function $R(\mathscr{H})$ for nonergodic information sources in terms of an average across the ergodic components of that source (e.g., Shields et al. 1978; Leon-Garcia et al. 1979; Effros et al. 1994, etc.). This allows us to construct another Boltzmann pseudoprobability in the resource rate Z as

$$dP(R, Z) = \frac{\exp[-R/g(Z)]dR}{\int_0^\infty \exp[-R/g(Z)]dR} \tag{2.12}$$

Again, we define a free energy analog as in terms of the 'partition function' giving

$$\exp[-F/g(Z)] = \int_0^\infty \exp[-R/g(Z)]dR = g(Z)$$
$$F(Z) = -g(Z)\log[g(Z)]$$
$$g(Z) = \frac{-F(Z)}{W(n, -F(Z))} \tag{2.13}$$

leading towards another version of the previous results. Again, $W(n, x)$ is the Lambert W-function of order n.

References

Atlan H, Cohen I (1998) Immune information, self-organization, and meaning. Int Immunol 10:711–717

Brown R (1992) Out of line. R Inst Proc 64:207–243

Cayron C (2006) Groupoid of orientational variants. Acta Crystalographica Sect A A62:21040

Champagnat N, Ferriere R, Meleard S (2006) Unifying evolutionary dynamics: from individual stochastic process to macroscopic models. Theor Popul Biol 69:297–321

Cover T, Thomas J (2006) Elements of information theory, 2nd edn. Wiley, New York

de Groot S, Mazur P (1984) Nonequilibrium thermodynamics. Dover, New York

Dembo A, Zeitouni O (1998) Large deviations and applications, 2nd edn. Springer, New York

Effros M, Chou A, Gray R (1994) Variable-rate source coding theorems for stationary nonergodic sources. IEEE Trans Inf Theory 40:1920–1925

Feynman R (2000) Lectures on computation. Westview Press, New York

Gould S, Lewontin R (1979) The spandrels of San Marco and the Panglossian paradigm: a critique of the adaptationist programme. In: Proceedings of the Royal Society of London, B, 205:581–598

Gray R (1988) Probability, random processes, and ergodic properties. Springer, New York

Jackson D, Kempf A, Morales A (2017) A robust generalization of the Legendre transform for QFT. J Phys A 50:225201

Khinchin A (1957) Mathematical foundations of information theory. Dover, New York

Laidler K (1987) Chemical kinetics, 3rd edn. Harper and Row, New York

Leon-Garcia A, Davisson L, Neuhoff D (1979) New results on coding of stationary nonergodic sources. IEEE Trans Inf Theory IT-25:137–144

Maignan A, Scott T (2016) Fleshing out the generalized Lambert W function. ACM Commun Comput Algebr 50:45–60

Matsumoto Y (1997) An introduction to Morse theory. American Mathematical Society, Providence, RI

Mezo I, Keady G (2015) Some physical applications of generalized Lambert functions. arXiv:1505.01555v2 [math.CA] 22 Jun 2015

Newman M (2010) Networks: an introduction. Oxford University Press, New York

Pettini M (2007) Geometry and topology in Hamiltonian dynamics and statistical mechanics. Springer, New York

Schreiber U, Skoda Z (2010) Categorified symmetries. arXiv:1004.2472v1

Scott T, Mann R, Martinez RE (2006) General relativity and quantum mechanics: towards a generalization of the Lambert W function. Appl Algebr Eng Commun Comput 17:41–47

Shields P, Neuhoff L, Davisson D, Ledrappier F (1978) The distortion-rate function for nonergodic sources. Ann Probab 6:138–143

Tateishi A, Hanel R, Thurner S (2013) The transformation groupoid structure of the q-Gaussian family. Phys Lett A 377:1804–1809

Tegmark M (2016) Improved measures of integrated information. PLOS Comput Biol. https://doi.org/10.1371/journal.pcbi.1005123

Wallace R (2005) Consciousness: a mathematical treatment of the global neuronal workspace model. Springer, New York

Wallace R (2012) Consciousness, crosstalk, and the mereological fallacy: an evolutionary perspective. Phys Life Rev 9:426–453

Wallace R (2017) Computational psychiatry: a systems biology approach to the epigenetics of mental disorders. Springer, New York

Wallace R (2018) New statistical models of nonergodic cognitive systems and their pathologies. J Theor Biol 436:72–78

Wallace R (2020) Signal transduction in cognitive systems: origin and dynamics of the inverted-U/U dose-response relation. J Theor Biol 504. https://doi.org/10.1016/j.jtbi.2020.110377

Wallace R (2021) Toward a formal theory of embodied cognition. Biosystems. https://doi.org/10.1016/j.biosystems.2021.104356

Weinstein A (1996) Groupoids: unifying internal and external symmetry. Not Am Math Assoc 43:744–752

Winkelbauer K (1970) On the asymptotic rate of non-ergodic information sources. Kybernetika Cislo 2. Rocnik 6/1970:127–148

Chapter 3
Examples and Extensions

3.1 Introduction

We have assembled enough tools for a simple application, recognizing the basic punctuated transition in consciousness—on/off—as being roughly analogous to the Data Rate Theorem in that there is a critical value of information flow rate H_0 for full function. We will have more to say on this in Chap. 4. That is, allowing a continuous approximation to the sum in the second expression of Eq. (2.6), we assume a minimum necessary critical limit H_0 and can write

$$\exp[-F/g(Z)] = \int_{-H_0}^{\infty} \exp[-(H_0 + x)/g(Z)]dx = g(Z)$$

$$F = -\log[g(Z)]g(Z)$$

$$g(Z) = \frac{-F(Z)}{W(n, -F(Z))} \tag{3.1}$$

where $W(n, x)$ is the Lambert W-function of order n that satisfies the relation $W(n, x) \exp[W(n, x)] = x$. It is real-valued only for orders $n = 0, -1$ over respective ranges $-\exp[-1] < x < \infty$ and $-\exp[-1] < x < 0$.

The appearance of the Lambert W-function is a distinct red flag, implying the possibility of re-envisioning and reconstructing the underlying problem in terms of a more fundamental, if substantially more abstract, submodular network. See Newman (2010) for general arguments and examples, and Yi et al. (2011) for an application to neural networks with time delays. Recall that the fraction of nodes in the 'giant component' within a random network of N nodes can be described in terms of the probability of contact between nodes p as $\{W(0, -Np \exp[-Np]) + Np\}/Np$. This expression is highly punctuated in the variable Np, leading, as discussed in the next chapter, to an elementary model of the accession to consciousness in a tunable linked system of cognitive submodules.

R. Wallace, *Consciousness, Cognition and Crosstalk: The Evolutionary Exaptation of Nonergodic Groupoid Symmetry-Breaking*, SpringerBriefs in Computational Intelligence, https://doi.org/10.1007/978-3-030-87219-9_3

Equation (3.1) leads to an expression for the cognition rate—in an argument abducted from chemical reaction theory (Laidler 1987)—as

$$L(Z) = \frac{\int_{H_0}^{\infty} \exp[-x/g(Z)]dx}{\int_{-H_0}^{\infty} \exp[-(H_0+x)/g(Z)]dx} = \exp[-H_0/g(Z)] \qquad (3.2)$$

Thus, in Eq. (2.6), we find $h(g(Z)) = g(Z)$ and can carry out an explicit calculation for g in terms of $f(Z)$ from Eq. (2.7), giving

$$g(Z) = \frac{-C_1 Z - Z \int \frac{f(Z)}{Z} dZ + C_2 + \int f(Z) dZ}{W(-C_1 Z - Z \int \frac{f(Z)}{Z} dZ + C_2 + \int f(Z) dZ)} \qquad (3.3)$$

where, again, $dZ/dt = f(Z(t))$ defines the adaptation function f, and $W(x)$ is the Lambert W-function, taken here of of order 0 and real-valued for $-\exp[-1] < x < \infty$.

It is important to recognize that, in higher animals, metabolic free energy is provided by the hydrolysis of adenosine triphosphate (ATP) to diphosphate (ADP). The free energy available from this reaction is significant, ranging between 30-60 KJ/mol, dependent on embedding physiological details. The energy supplied by this process can be equivalent to thousands of degrees K, suggesting that neural tissues, which typically consume metabolic free energy at a rate ten times that of more ordinary tissues, can indeed be driven to operate at very high 'reaction rates'.

3.2   Arousal

Following Wallace (2021a), the Yerkes-Dodson law (Diamond et al. 2007) relates complex task performance to arousal for individual animals under experimental conditions. Figure 3.1, as adapted from Diamond et al. (2007), shows that, depending on the difficulty of the task, there can be either a 'topping out' or an inverted-U pattern for performance versus arousal.

Typically, $f(Z)$, the adaptation function, might be seen as taking an 'exponential' form, i.e., $f(Z(t)) = \beta - \alpha Z(t)$.

In Fig. 3.2, we plot two expressions for $L(Z)$, the cognition rate, from Eq. (3.2), letting $H_0 = 1$. Here, $\alpha = 0.1$ and $\alpha = 1.0$, while the arousal index, β, varies. $C_1 = -3$ for both, while C_2 takes the values -18 and -3. Here, the lighter curve, corresponding to $\alpha = 1.0$, represents the easier task.

With proper manipulation of parameters and boundary conditions, the formalism produces something remarkably like the Yerkes-Dodson law.

Figure 3.3, by contrast, examines the efficiency of cognition, $L(Z)/Z$, as a function of the arousal β. Hard problems are are, in this model, clearly far more demanding of resources than simple ones, suggesting that it is more efficient to break up a hard problem into a series of simpler ones, or, perhaps more to the point, to a parallel set of them.

Fig. 3.1 The canonical forms of the Yerkes-Dodson law, for simple and difficult tasks

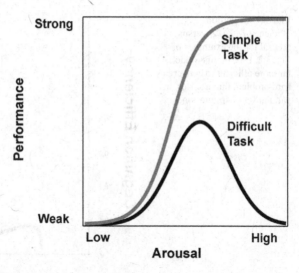

Fig. 3.2 Cognition rates $L(Z)$ versus β as an arousal index for simple and difficult tasks. Here, $H_0 = 1$, taking $\alpha = 0.1$ for the difficult and $\alpha = 1.0$, for the easier task. $C_1 = -3$ for both, while C_2 takes the values -18 and -3. The lighter curve, having $\alpha = 1.0$, represents the easier task

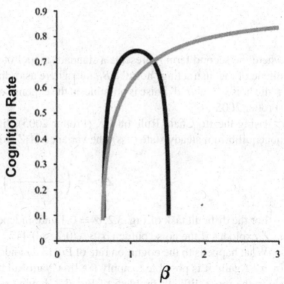

3.3 Distraction

Conscious effort and attention is not only affected by arousal, but by distraction, and it is not difficult to explore the dynamics of cognition rate $L(Z)$ using formalism available from the standard theory of stochastic differential equations (Protter 2005).

This requires expanding the relation $dZ/dt = f(Z(t)) = \beta - \alpha Z(t)$ as

$$dZ_t = (\beta - \alpha Z_t)dt + \sigma Z_t dB_t \tag{3.4}$$

Fig. 3.3 Cognition
efficiency $L(Z)/Z$ versus
arousal for the examples of
Fig. 3.2. It is, in this model,
far more efficient to convert a
hard problem into a series or
parallel set of simpler ones

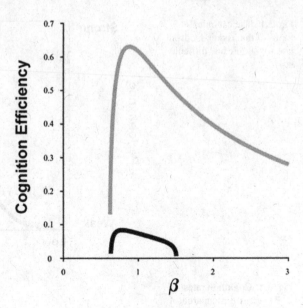

where the second term represents a standard model of 'volatility', with σ the magnitude of the distracting 'noise' dB_t, taken here as ordinary flat-spectrum Brownian white noise. 'Colored' noise is possible, at the expense of mathematical complication (Protter 2005).

Using the Ito Chain Rule on Z_t^2, (Protter 2005), 'it is easy to show' that, for a nonequilibrium steady state (nss), the variance of Z_t for the exponential model is

$$< Z^2 > - < Z >^2 = \left(\frac{\beta}{\alpha - \sigma^2/2} \right)^2 - \left(\frac{\beta}{\alpha} \right)^2 \tag{3.5}$$

For the difficult task of Fig. 3.2, $\alpha = 0.1$ and, independent of arousal β, variance in Z explodes if the noise burden $\sigma > \sqrt{0.2} \approx 0.447$.

What happens to the cognition rate of Eq. (3.2) under noise/distraction measured by σ? Again, it is possible to apply the Ito Chain Rule to $L(Z)$ as it was to Z^2. We study the more difficult problem of Fig. 3.2, having $\alpha = 0.1$, but fix $\beta = 1.1$, i.e., at the peak value of the cognition rate L. The resulting relation at nss—the solution set $\{\sigma, Z\}$ to the equation $< dL_t >= 0$—is literally too long to write on this page, but can be solved numerically via the *implicitplot* function of the computer algebra program Maple 2020, giving Fig. 3.4.

The cognition rate $L(Z)$ is very highly sensitive to the magnitude of distraction σ in this model. While, for the more difficult problem of Fig. 3.2, Z itself becomes unstable if $\sigma > 0.447$, L faces the possibility of a bifurcation instability for any $\sigma > 0$, and collapses entirely if $\sigma > 0.07$.

In the Mathematical Appendix we explore an intermediate example, studying the effect of 'distraction' σ on $F = -\log[g(Z)]g(Z)$ rather than on $L =$

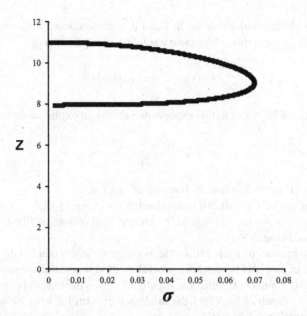

Fig. 3.4 Application of the Ito Chain Rule to the cognition rate $L(Z)$, with $f(Z) = 1.1 - 0.1Z$. The value of β has been taken as the peak for the more difficult problem of Fig. 3.2. We find the solution set $\{\sigma, Z\}$ for the nss relation $< dL_t >= 0$ using numerical methods. Instability begins at a much lower value of σ than 0.447, the limit for variance in Z as driven by simple volatility for $\alpha = 0.1$. By contrast, L faces the possibility of a bifurcation instability for any $\sigma > 0$, and fully collapses under the burden of distraction if $\sigma > 0.07$

$\exp[-H_0/g(Z)]$, again taking $dZ/dt = f(Z) = \beta - \alpha Z$. The relation $< dF_t >= 0$ is exactly solvable, via Lambert W-functions, and the appearance of such functions having orders -1 and 0 leads to an explicit punctuated bifurcation in σ.

3.4 Distraction and Arousal Under Fixed Delay

Rudolph and Repenning (2002) describe the underlying problems of delay and distraction in organizations as follows:

> ...[A]n overaccumulation of interruptions can shift an organizational system from a resilient, self-regulating regime, which offsets the effects of this accumulation, to a fragile, self-escalating regime that amplifies them.

Here, we explore general models of this phenomenon, across the full spectrum of cognitive enterprise.

Imposition of a fixed, discrete, delay δt on the adaptation function $f(Z) = \beta - \alpha Z(t)$, so that

$$dZ(t) = \beta - \alpha Z(t - \delta t) \qquad (3.6)$$

gives a delay-differential equation. In general, these are most difficult to analyze. Here, however, we can directly impose a solution to Eq. (3.6) as

$$Z_s(t) \equiv \frac{\beta}{\alpha}(1 - \exp[st]) \tag{3.7}$$

Application of Eq. (3.6) to this expression permits an explicit solution for s:

$$s = \frac{W(n, -\alpha \delta t)}{\delta t} \tag{3.8}$$

where, again, W is the Lambert W-function of order n.

Recall that, only for $n = 0$, is it real-valued, for $-\exp[-1] < x < \infty$, and for $n = -1$, real-valued for $-\exp[-1] < x < 0$. Lambert W-functions of all other orders are complex-valued (Fig. 3.5).

$\alpha \times \delta t$, the product of a rate and a time, is a dimensionless number driving system dynamics. Yi et al. (2011) extend the general method to multidimensional systems, using the matrix Lambert W-function. Figure 3.5 shows the real and complex values of the $n = -1$ branch of Eq. (3.8), the fundamental form for what we will do here, as functions of the product $\alpha \delta t$.

There are two critical values for the (dimensionless) index $\alpha \delta t$, derived from the appearance of the Lambert W-function, here taken of order -1. The first is at the point where the complex component becomes nonzero, i.e., when $\alpha \delta t > \exp[-1]$. This signifies onset of dying oscillatory dynamics. The second critical value is the value of $\alpha \delta t$ at which the real component of s becomes greater than zero, triggering explosive growth in oscillations. Note that the periodicity, determined by the magnitude of the complex component, changes as $\alpha \delta t$ increases beyond the first critical point.

Equation (3.7) implies

$$dZ_s/dt = s(Z_s - \beta/\alpha) \equiv f(Z_s) \tag{3.9}$$

permitting, as above, analysis of the resource delivery system's properties under stochastic fog, via the stochastic differential equation

$$dZ_t^s = s(Z_t^s - \beta/\alpha)dt + \sigma Z_t^s dB_t \tag{3.10}$$

Again, the second term represents volatility under conditions of Brownian noise. Applying the Ito Chain Rule to Z_s^2 gives the variance as

$$<Z_s^2> - <Z_s>^2 = \left(\frac{s\beta/\alpha}{s + \sigma^2/2}\right)^2 - \left(\frac{\alpha}{\beta}\right)^2 \tag{3.11}$$

Again, the system becomes grossly unstable if $\alpha \delta t > \exp[-1]$. Thus delay—$\delta t$— can greatly exacerbate inherent stochastic instabilities.

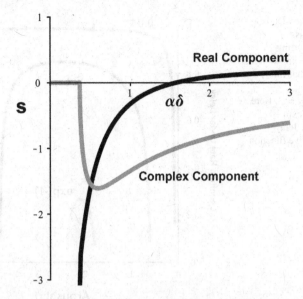

Fig. 3.5 s from Eq. (3.8), taking $n = -1$, real-valued only for $-exp[-1] < -\alpha\delta t < 0$. There are two critical values for the dimensionless index $\alpha\delta t$, derived from the appearance of the Lambert W-function. The first is at the point where the complex component becomes nonzero, representing the onset of dying oscillatory dynamics when $\alpha\delta t > exp[-1]$. The second is the point at which the real component of s becomes greater than zero, implying explosive growth in oscillations. Periodicity, determined by the magnitude of the complex component, changes as $\alpha\delta t$ increases beyond the first critical point

However, even if s is real-valued and negative, sufficient noise, measured by $\sigma^2/2$, also triggers explosive instability.

Recall the expressions for $g(Z)$ and $L(Z)$ from Eqs. (3.2) and (3.3), assuming a delayed adaptation function, so that $\partial Z_s/\partial t = f(Z_s(t)) = s(Z_s(t) - \beta/\alpha)$, where s is from Eq. (3.8), so that $Z_s(t) \to \beta/\alpha$.

Numerical exploration finds instability in L can be imposed by even very small delays δt.

We construct another Yerkes-Dodson 'arousal' analysis, showing cognition rate as a function of β, letting $Z = \beta/\alpha$ and taking appropriate values for other parameters. This is done in Fig. 3.6, where, for the stable—real-valued only—solutions, we set $\alpha = 1$, $C_1 = C_2 = 3$ and plot $L(\beta)$ for $\delta t = 0.04$, 0.3, $exp[-1]$, taking the Lambert W-function of order -1 for s in Eq. (3.19) and 0 in the expression for $g(Z_s)$. As the fixed delay δt increases, the 'inverted-U' of institutional cognition progressively collapses in a manner consistent with Delayed Auditory Feedback studies in which an artificially-induced delay of about 175ms between speech and hearing triggers extreme stress.

Fig. 3.6 A Yerkes-Dodson
'arousal' analysis for
cognition rate, setting
$n = -1$ in the expression for
s and $n = 0$ in that for g,
with $Z = \beta/\alpha$,
$\alpha = 1$, $C_1 = C_2 = 3$. Here,
$\delta t = 0.04$, 0.3, $\exp[-1]$.
Increasing fixed delay
collapses cognitive function

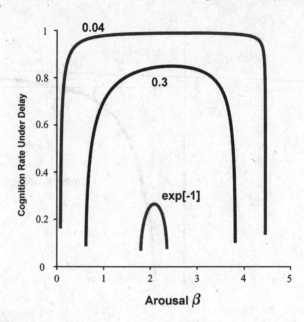

3.5 Two-Mode Dynamics

Next, we examine what is perhaps the simplest possible example, a two-mode non-ergodic system for which the high probability meaningful sequences are assumed to fall into two sets, each taken to be of the same size N, having source uncertainties $H_\pm = H_0 \pm \delta$ for a fixed $\delta > 0$. The larger value represents the 'on' mode, and the smaller the 'off'. Here, we—in effect—invoke the Ergodic Decomposition Theorem, as the $H_\pm$ represent ergodic 'extreme points' across the full nonergodic regime.

Some thought gives

$$\exp[-F/g(Z)] = N \exp[-H_0/g(Z)]2\cosh(\delta/g(Z))$$

$$F = -\log[2N\cosh(\delta/g(Z))]g(Z) + H_0$$

$$L(Z) = \frac{\exp[-\delta/g(Z)]}{2\cosh(\delta/g(Z))} \qquad (3.12)$$

where $L(Z)$ is again the cognition rate.

Again, we impose a first-order Onsager model, taking $S \equiv -F + Z\,dF/dZ$ and $\partial Z/\partial t \propto dS/dZ = f(Z)$.

Approximating the resulting relations to fourth order in δ gives, surprisingly, a second order expression as

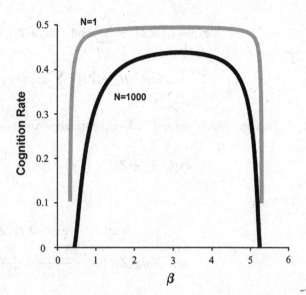

Fig. 3.7 Yerkes-Dodson arousal plot approximation, showing cognition rate L versus β for the adaptation function $f(Z) = \beta - \alpha Z$. Here, $N = 1, 1000$, $\delta = 0.1, \alpha = 1, Z = \beta$ and the boundary conditions are $C_1 = -2, C_1 = -1$

$$-Z\left(\ln(2) + \ln(N)\right)\left(\frac{d^2}{dZ^2}g(Z)\right) + Z\left(\frac{\frac{d^2}{dZ^2}g(Z)}{2g(Z)^2} - \frac{\left(\frac{d}{dZ}g(Z)\right)^2}{g(Z)^3}\right)\delta^2 \approx f(Z)$$

(3.13)

We will again take $f(Z) = \beta - \alpha Z$. The resulting equation can be explicitly solved for $g(Z)$, leading to a distractingly complicated expression for the cognition rate which we omit for clarity.

Figure 3.7, however, shows the resulting—and highly approximate—Yerkes-Dodson arousal relations, i.e., the $L(\beta)$, for two different values of N. Here, $\alpha = 1, Z = \beta, N = 1000, 1, \delta = 0.1$, with the necessary two boundary conditions as $C_1 = -2, C_2 = -1$.

Because of the simple structure—only two modes—we have been able to explicitly derive the h-function relation of Eqs. (2.4) and (2.6).

3.6 Multi-mode Dynamics

Suppose there a many, say N, possible H-values, clustered around a base value H_0. Then

$$\exp[-F/g(Z)] = \sum_{j=1}^{N} \exp[-(H_0 + \delta_j)/g(Z)] =$$

$$\exp[-H_0/g(Z)] \left(\sum_{j=1}^{N} \exp[-\delta_j/g(Z)] \right) \qquad (3.14)$$

We can estimate the sum in δ_j using the approximation

$$\exp[-\delta_j/g(Z)] \approx 1 - \delta_j/g(Z) + \frac{1}{2} \frac{\delta_j^2}{g(Z)^2} \qquad (3.15)$$

Then

$$\exp[-F/g(Z)] \approx$$
$$\exp[-H_0/g(Z)] \left(N - \frac{\sum_j \delta_j}{g(Z)} + \frac{\sum_j \delta_j^2}{2g(Z)^2} \right) \qquad (3.16)$$

so that

$$F(Z) \approx - \log \left(N + \frac{\Delta}{2g(Z)^2} \right) g(Z) + H_0 \qquad (3.17)$$

where $\Delta \equiv \sum_j \delta_j^2$ and we have adjusted H_0 so that $\sum_j \delta_j = 0$.

We again introduce an entropy-analog as $S = -F + Z dF/dZ$, and impose Onsager dynamics as $\partial S/\partial t \propto dS/dZ = f(Z)$.

Taking $f(Z) = \beta - \alpha Z$, and assuming $\Delta/2g(Z)^2 \gg N$ leads to

$$g(Z) = \frac{2 \ln(Z) Z\beta - \alpha Z^2 + 2C_1 Z - 2\beta Z - 2C_2}{4W \left(n, -\frac{\sqrt{\frac{\left(2\ln(Z)Z\beta - \alpha Z^2 + 2C_1 Z - 2\beta Z - 2C_2\right)^2}{\Delta}}}{4} \sqrt{2} \right)} \qquad (3.18)$$

where, again, $W(n, x)$ is the Lambert W-function of order n.

Recall that $W(n, x)$ is real-valued only for $n = 0, -1$ and only over limited ranges in x, suggesting punctuated phase transitions in this expression, driven by symmetry-breaking in the groupoids defining the real-value ranges of x.

The model leading to Eq. (3.18) is obviously fragile, but, assuming a Data Rate Theorem limit H_1 for a cognition rate defined as $L \propto \exp[-H_1/g(Z)]$, proper choice of boundary conditions does indeed produce the ubiquitous inverted-U. In Fig. 3.8, $H_1 = 1, n = -1, \Delta = 1000, \alpha = 1, C_1 = -3, C_2 = 1$, assuming $Z = \beta/\alpha$. The approximation fails for large β.

Again following Wallace (2020b), another possible distributed state model takes states as evenly dispersed about the DRT limit H_0 within limits $\pm\varepsilon > 0$ so that the partition function and cognition rate relations become

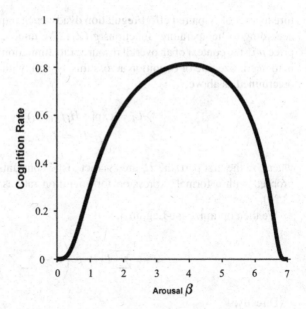

Fig. 3.8 Cognition rate example for Eq. (3.18). Here, $L \propto \exp[-H_1/g(Z)]$, setting $H_1 = 1$, $n = -1$, $\Delta = 1000$, $\alpha = 1$, $C_1 = -3$, $C_2 = 1$, $Z = \beta/\alpha$. The approximation is fragile and fails at larger β

$$\int_{-\varepsilon}^{\varepsilon} \exp[-(H_0 + x)/g(Z)]dx =$$

$$g(Z)\,(\exp[(-H_0 + \varepsilon)/g(Z)] - \exp[(-H_0 - \varepsilon)/g(Z)])$$

$$\equiv \exp[-F(Z)/g(Z)]$$

$$L(Z) = \int_0^{\varepsilon} \exp[-(H_0 + x)/g(Z)]dx =$$

$$g(Z)\,(\exp[-H_0/g(Z)] - \exp[(-H_0 - \varepsilon)/g(Z)]) \qquad (3.19)$$

Carrying through the Onsager dynamics based on the gradient in Z of $S(Z) = -F(Z) + Z\,dF/dZ$ requires expanding the resulting equation as a fifth order series in ε (surprisingly, giving an equation in second order involving ε itself). Taking $\partial Z/\partial t = dS/dZ = f(Z) = \beta - \alpha Z(t)$ produces a straightforward but complicated relation for $g(Z)$, again depending on two boundary conditions. This expansion gives results for rates of cognition similar to those explored above.

3.7 Environmentally-Induced Cognitive Failure

We have argued that consciousness, particularly in view of its stripped-down, speeded-up character, is inherently unstable and requires significant effort to keep the 'stream of consciousness' within regulatory riverbanks. We can extend this picture to a set of N paired effort/regulation dyads, each requiring resources at rates Z_i according to 'temperature' functions $g_i(Z_i)$. We impose an environmental shadow price μ in the context of an overall resource rate limitation $Z = \sum_i Z_i$. One approach is to maximize rate of cognition across this system, with each level's cognition rate determined as above:

$$L_i = \exp[-H_i/g_i(Z_i)]$$
$$L_i = L_i(Z_i) \tag{3.20}$$

where, in the first part, the H_i are characteristic constants. However, we can simply proceed with a formal expression for cognition rate as in the second part of Eq. (3.20).

We then optimize the Lagrangian

$$\mathcal{L} = \sum_i L_i + \mu \left(Z - \sum_i Z_i \right) \tag{3.21}$$

Directly,

$$\partial \mathcal{L}/\partial Z_i = 0$$
$$dL_i(Z_i)/dZ_i = \mu \tag{3.22}$$

so that, across any functional form for the cognition rate $L(Z)$, or any other other 'quality' measure,

$$L_i = \mu\,(Z_i + C_i) \tag{3.23}$$

where the C_i are boundary conditions.

Since $Z_i < Z \; \forall \; i$, and since we may expect the C_i to be relatively small, proper choice of shadow price μ can drive one or more critical cognition rates below levels needed for effective system function.

If there is, in addition, a time constraint, so that

$$\mathcal{L} = \sum_i L_i + \mu \left(Z - \sum_i Z_i \right) + \lambda \left(T - \sum_i T_i \right) \tag{3.24}$$

then some further calculation finds, across any functional form for L_i,

$$Z_i(T_i) = \frac{\lambda}{\mu} T_i + C_i \tag{3.25}$$

to much the same effect, since $T_i < T$, here driven by the shadow price ratio.

Such dynamics are of central interest, not only in the study of physiological and psychological pathologies (Wallace and Wallace 2016; Wallace 2017), but also in understanding and directing organized conflict (Wallace 2018, 2019, 2020a).

3.8 Optimization Under Delay and Resource Constraint

We suppose a multilevel, multicomponent organism (or other distributed cognitive/conscious entity) is constrained by both time and resource rate, taking the time dependences of resource delivery as either discrete or exponentially distributed, so that, for each level,

$$dZ/dt = s(Z - \beta/\alpha)$$
$$dZ/dt = \beta - \alpha \int_0^t Z(t - \tau) m \exp[-m\tau] d\tau \tag{3.26}$$

as appropriately indexed.

We then seek to optimize a Lagrangian $\mathscr{L}$ as

$$\mathscr{L} = Q[Z_1(T_1), ..., Z_n(T_n)] + \lambda \left(T - \sum_j T_j \right) + \mu \left(Z - \sum_j Z_j \right)$$
$$\partial \mathscr{L}/\partial Z_j = \partial Q/\partial Z_j - \mu = 0$$
$$\partial \mathscr{L}/\partial T_j = (\partial Q/\partial Z_j) dZ_j/dT_j - \lambda = \mu \, dZ_j/dT_j - \lambda = 0 \tag{3.27}$$

where, as above, $Q[Z_1(T_1), ..., Z_j(T_n)]$ is some generalized 'quality' measure.

We are led to the relations

$$\lambda/\mu = s_j(Z_j - \beta_j/\alpha_j)$$
$$\lambda/\mu = \beta_j - \alpha_j \int_0^{T_j} Z_j(T_j - \tau) m_j \exp[-m_j\tau] d\tau \tag{3.28}$$

Solving—for the the second using the Laplace transform—gives

$$Z_j(T_j) = \frac{\beta_j}{\alpha_j} + \frac{1}{s_j} \frac{\lambda}{\mu}$$
$$Z_j(T_j) = \frac{\beta_j}{\alpha_j} - \frac{1}{\alpha_j} \frac{\lambda}{\mu} \tag{3.29}$$

The 'undetermined multipliers' λ and μ are to be interpreted from economic theory, i.e., as 'shadow prices' in the sense of Jin et al. (2008) that are imposed on optimization by 'environmental externalities', for example, imminent threat from a predator. Recall that, for stability in Fig. 3.5, the s_j must all be negative.

One may, of course, impose other distributions for the delay.

The implication of Eq. (3.29) is that the shadow price ratio λ/μ—characterizing the influence of externalities—becomes a powerful selection pressure determining system failure or success. Too high a level of threat drives the Z_j below rates needed for successful function.

3.9 A Remark on Multiple Delays

Equation (3.6) assumes a single delay in resource delivery. There may, however, be many such, so that we must write $dZ/dt = \beta - \sum_{j=1}^{N} \alpha_j Z(t - \delta t_j)$

Some thought finds it possible to impose a solution as

$$Z(t) = \frac{\beta}{\sum_{j=1}^{N} \alpha_j} (1 - \exp[st]) \qquad (3.30)$$

leading to an implicit relation defining s as

$$s = -\sum_{j=1}^{N} \frac{\alpha_j}{\exp[s\delta t_j]} \qquad (3.31)$$

Further development shows that s has a negative real part only over a limited region of the δt_j near zero.

For distributed delays,

$$dZ/dt = \beta - \sum_{j=1}^{N} \alpha_j \int_0^t Z(t - \tau) m_j \exp[-m_j \tau] d\tau \qquad (3.32)$$

leading to an expression for the Laplace transform of Z as

$$r\mathscr{L}(Z(t), t, r) = \frac{\beta}{r} - \left(\sum_{j=1}^{N} \frac{\alpha_j m_j}{r + m_j} \right) \mathscr{L}(Z(t), t, r) \qquad (3.33)$$

Solving for, and then inverting $\mathscr{L}(Z(t), t, r)$, gives $Z(t)$ as a sum across terms in $\exp[Q_j t]$, where the Q_j are the solutions to a polynomial of degree $N + 1$ in the α_j and the m_j. The Q_j have negative real parts only for sufficiently large cognition rates m_j.

3.10 Network Topology and System Cognition Rate

Expanding remarks in Wallace (2020b), suppose there is a networked system of n interacting cognitive components, each with its own rate of essential resources Z_q and cognitive temperature $g_q(Z_q)$, with a vector of resource rates written as $\mathbf{Z} = (Z_1, ..., Z_n)$, $Z_q \geq 0$ that may possibly be constrained as $Z = \sum_q Z_q$. Each subsystem will operate at some cognition rate $L_q = \exp[-H_q/g_q(Z_q)]$. The way in which the full system operates depends critically on the underlying network topology, as well as on resource constraints.

- If there is a single, highly dominant, bottleneck at a particular level C, then

$$L_{sys} \approx \exp[-H_C/G_C(Z_C)] \tag{3.34}$$

- For a simple chain-effect model

$$L_{sys} \equiv \exp[-1/G(\mathbf{Z})] = \Pi_q \exp[-H_q/g_q(Z_q)]$$
$$1/G(\mathbf{Z}) = \sum_q H_q/g_q(Z_q) \tag{3.35}$$

- For a highly parallel system,

$$L_{sys} = \exp[1/G(\mathbf{Z})] = \sum_q \exp[-H_q/g_q(Z_q)]$$
$$1/G(\mathbf{Z}) = -\log\left(\sum_q \exp[-H_q/g_q(Z_q)]\right) \tag{3.36}$$

All such relations may possibly be subject to a total resource rate constraint.

Real-world networks are likely to be far less tractable. For such systems no single cognitive temperature may be possible. Casas-Vazquez and Jou (2003) describe the problem in physical systems as follows,

> When the assumption of local equilibrium is no longer tenable... one is faced with the problem of defining temperature and entropy in non-equilibrium conditions. This can occur for instance when the relaxation times of some internal degrees of freedom are long or when the values of the fluxes or gradients present in the system are high...

Real-world, real-time cognitive systems will seldom or ever achieve nonequilibrium steady state, and extension of the ideas presented here remains to be done.

3.11 Expanding the Onsager Approximation

Following Wallace (2021b), we can generalize the formalism in the direction indicated by Eq. (2.8), defining a new 'Onsager Approximation' as

$$\partial Z/\partial t \approx dS/dZ - \varepsilon d^2 S/dZ^2 = f(Z) \tag{3.37}$$

again constructing an entropy-analog from the iterated free energy F in terms of the Legendre Transform: $S \equiv -F(Z) + ZdF/dZ$.

This generates a relation in F as

$$Z\left(\frac{d^2}{dZ^2}F(Z)\right) - \varepsilon\left(\frac{d^2}{dZ^2}F(Z) + Z\left(\frac{d^3}{dZ^3}F(Z)\right)\right) = f(Z) \tag{3.38}$$

having the solution

$$F(Z) = \int\int -\frac{\left(-C_1\varepsilon + \int f(Z)\,e^{-\frac{Z}{\varepsilon}}dZ\right)e^{\frac{Z}{\varepsilon}}}{Z\varepsilon}dZdZ + C_2Z + C_3 \tag{3.39}$$

The C_j are boundary conditions.

The relation $F(Z) = -\log[h(g(Z))]g(Z)$ can then be solved for $g(Z)$.

Other higher order expressions, like those of Eq. (2.8), can often be explicitly (or implicitly) solved in a similar manner (Wallace 2021b). The centrality is the fitting of 'regression models' to observational or experimental data. Sometimes $Y = mX + b$ works, sometimes $Y = mX^2 + b$ works better, and so on. Scientific inference follows from what fits, what does not, and how good the fit may be. For complicated biological, psychological, and social phenomena, data structures, not mathematical models or statistical considerations, drive the science.

3.12 Reconsidering 'Delay'

We explored delay in terms of time alone. Environmental interactions may, however, be quite multidimensional, involving a convolution of space, time, and, in some sense, 'cogspace', on which diffusional constraints determine a maximum 'velocity' at which a signal can propagate. Recent work has treated brain function in terms of a relativistic pseudo-diffusion framework (Le Bihan 2020) that seems—with some considerable work—adaptable to what we have done here. To paraphrase that work, considering that the propagation speed of actions on an environmental landscape of space, cogspace and time, is limited by fog, friction, and cognitive intent, it seems possible to apply concepts borrowed from the physical theories of relativity to introduce the view that space, cogspace, and time are tightly blended in environmental interaction. It can be argued that physiological, social, institutional and other structural features might be unified through a combined 'cognitive spacetime'. This multidimensional construct presents a functional curvature generated by interaction that is roughly analogous to our four-dimensional universe spacetime in its curvature. The analogy may be extended to stochastic realms.

This approach requires defining a multidimensional metric something like

$$d\mathscr{S}^2 = c^2 dt^2 - \sum_j dx_j^2 \tag{3.40}$$

where the latter sum is across (social or geographic) space and 'cogspace' (scalarized in some appropriate sense) and c is the maximum rate at which 'resources' may actually be sent across an extended environment during interaction. The next step would involve exploring discrete and distributed delays in $\mathscr{S}$ over the relation

$$dH/d\mathscr{S} = \beta - \alpha H(\mathscr{S}) \tag{3.41}$$

Presumably, results similar to those of the previous sections emerge, in $\mathscr{S}$ rather than in t. There are, however, important subtleties. More generally, one defines $\mathscr{S}$ as

$$d\mathscr{S}^2 = \mathscr{G}_{\mu\nu} dx^\mu dx^\nu \tag{3.42}$$

using a summation convention for which $\mathscr{G}_{\mu\nu}$ is an appropriate metric tensor. Further development is standard, but arduous, leading into unexpectedly deep waters (e.g., Herrmann 2009; Castro Villarrea 2010).

3.13 A More Radical Program

Again following Wallace (2021b), it is possible to make further extension of theory, further expanding the idea of 'entropy' by fully generalizing the Legendre transform construct built from the iterated free energy Morse Function $F(Z)$.

The centrality of the classic Legendre transform is that it is self-inverse. Different generalizations can be carried out in more or less natural ways. For example, Jackson et al. (2017) characterize the Legendre transform in terms of formal power series:

> ..[T]he Legendre transform need not be viewed as a map from functions into functions or functionals into functionals. Instead, the Legendre transform can be viewed as mapping the coefficients of one formal power series into the coefficients of another formal power series. Here, the term 'formal' does not express 'mathematically non-rigorous', as it often does in the physics literature. Instead, the term 'formal power series' is here a technical mathematical term, meaning a power series in indeterminates. Formal power series are not functions. A priori, formal power series merely obey the axioms of a ring and questions of convergence do not arise.

For cognitive systems one need only impose dimensional consistency across such a formal series model, defining the entropy-analog as

$$S \equiv \sum_{n=1}^{\infty} \varepsilon_n Z^{n-1} F^{(n-1)}$$

$$\varepsilon_1 = -1, \; F^{(0)} = F(Z), \; F^{(k)} \equiv dF^k/dZ^k \tag{3.43}$$

Then a first order Onsager approximation becomes

$$\partial Z/\partial t \approx \partial S/\partial Z =$$

$$\sum_{n=1}^{\infty} \varepsilon_n (n-1) Z^{n-2} F^{(n-1)} +$$

$$\sum_{n=1}^{\infty} \varepsilon_n Z^{n-1} F^{(n)} = f(Z) \tag{3.44}$$

where, for algebraic completeness, one can likewise expand $f(Z)$ as a formal power series.

The 'epsilon spectrum' that determines the resulting ring algebra will likely be unique to a particular cognitive process at a particular time, requiring an understanding of 'epsilon dynamics' for full characterization.

References

Casas-Vazquez J, Jou D (2003) Temperature in non-equilibrium states: a review of open problems and current proposals. Rep Prog Phys 66:1937–2023

Castro Villarrea P (2010) Brownian motion meets Riemann curvature, arXiv:1005.0650v1

Diamond D, Campbell A, Park C, Halonen J, Zoladz P (2007) The temporal dynamics model of emotional memory processing... Neural Plast. https://doi.org/10.1155/2007/60803

Herrmann J (2009) Diffusion in the special theory of relativity. Phys Rev E 80:05110

Jackson D, Kempf A, Morales A (2017) A robust generalization of the Legendre transform for QFT. J Phys A 50:225201

Jin H, Hu Z, Zhou X (2008) A convex stochastic optimization problem arising from portfolio selection. Math Financ 18:171–183

Laidler K (1987) Chemical kinetics, 3rd edn. Harper and Row, New York

Le Bihan D (2020) On time and space in the brain: a relativistic pseudo-diffusion framework. Brain Multiphys 1:100016

Newman M (2010) Networks: an introduction. Oxford University Press, New York

Protter P (2005) Stochastic integration and differential equations: a new approach, 2nd edn. Springer, New York

Rudolph J, Repenning N (2002) Disaster dynamics: understanding the role of quantity in organizational collapse. Adm Sci Q 47:1–30

Wallace R, Wallace D (2016) Gene expression and its discontents: the social production of chronic disease, 2nd edn. Springer, New York

Wallace R (2017) Computational psychiatry: a systems biology approach to the epigenetics of mental disorders. Springer, New York

Wallace R (2018) Carl von Clausewitz, the Fog-of-War, and the AI revolution: the real world is not a game of Go. Springer, New York

Wallace R (2019) Contrasting tactical and strategic dynamics on a Clausewitz landscape. JDMS 17:143–153

Wallace R (2020a) Cognitive dynamics on Clausewitz landscapes: the control and directed evolution of organized conflict. Springer, New York

Wallace R (2020b) How AI founders on adversarial landscapes of fog and friction. JDMS. https://doi.org/10.1177/1548512920962227

Wallace R (2021a) Embodied cognition and its pathologies: the dynamics of institutional failure on wickedly hard problems. Commun Nonlinear Sci Numer Simul. https://doi.org/10.1016/j.cnsns.2020.105616

Wallace R (2021b) Toward a formal theory of embodied cognition. BioSystems 202:104356. https://doi.org/10.1016/j.biosystems.2021.104356

Yi S, Yu S, Kim JH (2011) Analysis of neural networks with time-delays using the Lambert W function. In: Proceedings of the 2011 American Control Conference, San Francisco, CA, USA, 2011, pp. 3221–3226. https://doi.org/10.1109/ACC.2011.5991085

Chapter 4
Phase Transitions

4.1 Introduction

We first examine the implications of 'no free lunch' and 'tuning theorem' arguments for the punctuated address of rapidly-changing patterns of threat and affordance facing a conscious organism.

Given a set of cognitive biological modules that become linked to solve a problem, for example riding a bicycle in heavy traffic, followed by wound healing, the 'no free lunch' arguments of Wolpert and MacReady (1997) become significant. The essential point (English 1996) is that if an optimizer has been tuned to the most effective possible structure for a particular kind of problem or problem set, it will necessarily be worst for some other problem set, which must then have a different function optimizer for optimality. As Wallace (2012) puts it,

> Another way of stating this conundrum is to say that a computed solution is simply the product of the information processing of a problem, and, by a very famous argument, information can never be gained simply by processing. Thus a problem X is transmitted as a message by an information processing channel, Y, a computing device, and recoded as an answer. By the 'tuning theorem' argument of the Mathematical Appendix, there will be a channel coding of Y which, when properly tuned, is itself most efficiently 'transmitted', in a sense, by the problem the 'message' X. In general, then, the most efficient coding of the transmission channel, that is, the best algorithm turning a problem into a solution, will necessarily be highly problem-specific. Thus there can be no best algorithm for all sets of problems, although there will likely be an optimal algorithm for any given set.

From these considerations, it becomes clear that different challenges facing an organism must be met by different arrangements of cooperating lower level cognitive modules. We make an abstract picture of this, not based on anatomy, but on the network of linkages between the information sources dual to the physiological and learned unconscious cognitive modules (UCM) that may become entrained into address of those challenges. The network of lower level cognitive modules is reexpressed in terms of the information sources dual to them. Given two distinct problem classes (e.g., bicycle riding vs. wound healing), there must be two markedly differ-

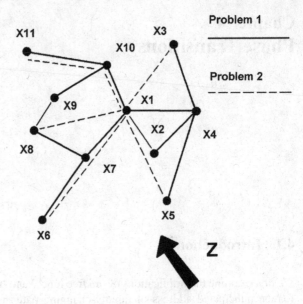

Fig. 4.1 Two different problems facing an organism will be optimally solved by two different linkages of available lower level unconscious cognitive modules (UCM), characterized here by their dual information sources X_j, into different temporary networks of working structures. These are linked by crosstalk among the sources rather than as the UCM themselves. The embedding information source Z represents the influence of external signals

ent wirings of the information sources dual to the available UCM, as in Fig. 4.1. The network graph edges are measured by the level of information crosstalk between sets of nodes representing the dual information sources.

The possible expansion of a closely-linked set of information sources dual to the UCM into a global broadcast depends, for this model, on the underlying network topology of the dual information sources and on the strength of the couplings between the individual components of that network.

We construct perhaps the simplest model of that process.

4.2 A First Model Class

The possible appearance of the Lambert W-function in the arguments above—for the simple case $h(g(Z)) = g(Z)$—is a warning. The fraction of nodes within the 'giant component' of a random network of N nodes—here, taken as interacting information sources dual to unconscious cognitive processes—can be described in terms of the probability of contact between nodes, p, as (Newman 2010)

$$\frac{W(0, -Np\exp[-Np]) + Np}{Np} \tag{4.1}$$

giving the results of Fig. 4.2.

Note, in particular, the threshold for highly punctuated onset of a single giant component in the random network case. This sort of dynamic is a central matter for

Fig. 4.2 Proportion of N interacting information sources dual to unconscious cognitive processes that are entrained into a 'giant component' global broadcast as a function of the probability of contact p for random and stars-of-stars-of-stars topologies. Tuning topologies determines the threshold for 'ignition' to a single, large-scale, global broadcast

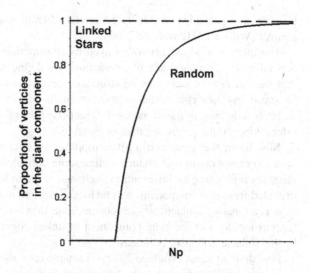

the high-speed neural processes that are the foundation of consciousness, seen here as a necessarily stripped-down example of more general multiple tunable spotlight physiological phenomena that can do much more, but only if they operate at far slower rates: gene expression, immune function, tumor suppression, and so on (Wallace 2005, 2012; Wallace and Wallace 2016).

An important feature here is the topological tunability of the threshold dynamics implied by the two limiting cases, the star-of-stars-of-stars versus the random network.

Lambert W-functions thus appear to suggest existence of an underlying formal network structure. For our purposes here—neural structures—we can envision the underlying abstract network to be a set of information sources dual to unconscious cognitive phenomena within the brain. These become linked by 'Np' crosstalk, in the context of a tunable topology that shifts somewhere between the two limits of the figure.

See Fig. 6 of Dehaene and Changeux (2011) for something similar.

Previous sections have abducted results from nonequilibrium thermodynamics to consciousness theory, applicable to nonergodic, as well as ergodic, models of cognition. Here, we abduct the Kadanoff renormalization treatment of physical phase transitions (e.g., Wilson 1971; Wallace 2005, 2012), applying it to a reduced version of the iterated 'free energy' Morse Function of Eq. (2.4), expanding the approach of Wallace (2021a).

Although a more general argument can be made, representing embodied consciousness *an sich*, for the sake of familiarity, we project down on to the subsystem dominated by $\mathscr{C}$, the internal system bandwidth, envisioning a number of internal cognitive submodules as connected into a topologically identifiable network having a variable average number of fixed-strength crosstalk linkages between components. The mutual information measure of crosstalk can continuously change, and it becomes then possible to conduct a parameterized renormalization in a now-standard manner (Wilson 1971; Wallace 2005).

The internal modular network linked by information exchange has a topology depending on the magnitude of interaction. We define an interaction parameter, a real number $\omega > 0$, and examine structures characterized in terms of linkages set to zero if crosstalk is less than ω, and renormalized to 1 if greater than or equal to ω. Each ω defines, in turn, a network 'giant component' (Spenser 2010), linked by information exchange greater than or equal to it.

Now invert the argument: a given topology of interacting submodules making up a giant component will, in turn, define some critical value ω_C such that network elements interacting by information exchange at a rate less than that value will be excluded from that component, will be locked out and not 'consciously' perceived.

ω is a tunable, syntactically dependent, detection limit depending on the instantaneous topology of the giant component of linked cognitive submodules defining, by that linkage, a 'global broadcast'.

For 'slow' systems (Wallace 2012)—immune response, gene expression, institutional process—as opposed to the 100 ms time constant of higher animal consciousness, there can be many such 'global workspace' spotlights acting simultaneously. Such multiple global broadcasts, indexed by the set $\Omega = \{\omega_1, \omega_2, \ldots\}$, lessen the likelihood of inattentional blindness to critical signals, both internal and external. The immune system, for example, engages simultaneously in pathogen and malignancy attack, neuroimmuno dialog, and routine tissue maintenance (Cohen 2000).

Assuming it possible to scalarize the set Ω in something of the manner of Z above, we work with a single, real-value ω, and can model the dynamics of a multiple tunable workspace system.

Recall the definition of the iterated free energy F from Eq. (2.4), now focused within and characterized by ω. The essential idea is to invoke a 'length' r on the network of internal interacting information sources. r will be more fully defined below. We follow the renormalization methodology of Wilson (1971) as described in Wallace (2005), although other approaches are clearly possible. That is, there is no unique renormalization symmetry.

The central idea is to invoke a 'clumping' transformation under an 'external field strength' that can be, in the limit, set to zero. For clumps of size R, given a field of strength J,

$$F[\omega(R), J(R)] = \mathscr{F}(R) F[\omega(1), J(1)]$$
$$\chi[\omega(R), J(R)] = \frac{\chi[\omega(1), J(1)]}{R} \qquad (4.2)$$

χ represents a correlation length across the linked information sources.

$\mathscr{F}(R)$ is a 'biological' renormalization relation that can take such forms as R^δ, $m \log(R) + 1$, $\exp[m(R - 1)/R]$, and so on, so long as $\mathscr{F}(1) = 1$ and is otherwise monotonic increasing. Physical theory is restricted to $\mathscr{F}(R) = R^3$.

Surprisingly, after some tedious algebra, the standard Wilson (1971) renormalization phase transition calculation drops right out for the extended relations, described first in Wallace (2005) and summarized in the Mathematical Appendix.

There remains a problem. Just what is the metric r? In this, we follow Wallace (2012).

First, impose a topology on the system of interacting information sources such that, near a particular 'language' A associated with some source uncertainty measure H, there is an open set U of closely similar languages $\hat{A}$ such that the set $A, \hat{A} \in U$.

Since the information sources are sufficiently similar, for all pairs of languages $A, \hat{A}$ in U it is possible to

- Create an embedding alphabet which includes all symbols allowed to both.
- Define an information-theoretic distortion measure in the extended joint alphabet between any high probability (i.e., properly grammatical and syntactical) paths in A and $\hat{A}$, written as $d(Ax, \hat{A}x)$. The different languages do not interact in this approximation.
- Define the metric on U as

$$r(A, \hat{A}) \equiv |\int_{A,\hat{A}} d(Ax, \hat{A}x) - \int_{A,A} d(Ax, A\hat{x})| \tag{4.3}$$

where Ax and $\hat{A}x$ are paths in the languages $A, \hat{A}$ respectively, d is the distortion measure, and the second term is a 'self-distance' for the language A such that $r(A, A) = 0, r(A, \hat{A}) > 0, A \neq \hat{A}$.

Some thought shows this version of r is sufficient, if somewhat counterintuitive. A more formal approach can be found in Glazebrook and Wallace (2009).

Extension of the Wilson technique to a fully-embodied consciousness model seems straightforward. However, since the dynamics of the embedded condition are so highly variable, there will be no unique solution, although there may well be equivalence classes of solutions, defining yet more groupoids in the sense of Tateishi et al. (2013).

Indeed, groupoids may rear their heads at a more fundamental level: Wilson's renormalization semigroup, in the cognitive circumstance of discrete equivalence classes of developmental pathways, might well require generalization as a renormalization semigroupoid, e.g., a disjoint union of different renormalization semigroups across a nested or otherwise linked set of information sources and/or iterated free energy constructs dual to cognitive modules. Something roughly analogous has been postulated for 'spin foam' gravity models (Oeckl 2003).

Following the arguments of Wallace (2018), we have, in effect, studied equivalence classes of directed homotopy developmental paths—the $\{x_1, \ldots, x_n, \ldots\}$—associated with nonergodic cognitive systems defined in terms of single-path source uncertainties. These require imposition of structure in terms of the metric r of Eq. (4.3), leading to groupoid symmetry-breaking transitions driven by changes in the temperature analog $g(Z)$. There can be an intermediate case under circumstances in which the standard ergodic decomposition of a stationary process is both reasonable and computable—no small constraint. Then there is an obvious natural directed homotopy partition in terms of the transitive components of the path-equivalence class groupoid. It appears that this decomposition is equivalent to, and maps on, the ergodic decomposition of the overall stationary cognitive process. It then becomes possible to define a constant source uncertainty on each transitive subcomponent, fully indexed by the embedding groupoid. This was done earlier for the two-mode example.

That is, each ergodic/transitive groupoid component of the ergodic decomposition recovers a constant value of the source uncertainty dual to cognition, presumably given by standard 'Shannon entropy' expression. Since it is possible to view the components themselves as constituting single paths in an appropriate quotient space, this leads to the previous 'nonergodic' developments.

As Wallace (2018) notes, the argument tends toward Mackey's theory of 'virtual groups', i.e., 'ergodic groupoids' (Hahn 1978; Mackey 1963; Series 1977).

A complication emerges through imposition of a double symmetry involving metric r-defined equivalence classes on this quotient space. That is, there are different possible strategies for any two teams playing the same game. In sum, however, groupoid symmetry-breaking in the iterated free energy construct of Eq. (2.4) or (2.10) will still be driven by changes in $g(Z)$ and/or ω.

4.3 A Second Model Class

The Data Rate Theorem (DRT), as instantiated in Eqs. (6.2) and (6.10), provides another perspective on the punctuated accession to consciousness of an environmental or other signal. The essential idea of the DRT is that a system that generates its own 'topological information' at some rate H_0 cannot be stabilized—here, made to respond to an external control signal—unless the control signal information source uncertainty H is greater than H_0, the rate at which the system of interest is taking care of its own business. That is, the 'control' signal is characterized in terms of the Z variate, representing a signal that must break through ongoing business-as-usual to achieve accession to consciousness.

Following Wallace (2021b), the 'obvious' extension of the DRT is the inequality

$$H(Z) > h(Z)H_0 \tag{4.4}$$

where, again, Z represents the influence of the intrusive signal that is attempting accession to consciousness.

Following a classic Black–Scholes argument (e.g., Wallace 2020, Sect. 14.4), an exactly solvable first approximation leads to

$$H(Z) \approx \kappa_1 Z + \kappa_2 \tag{4.5}$$

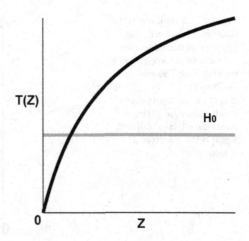

Expanding $h(Z)$ in first order, so that $h(Z) \approx \kappa_3 Z + \kappa_4$, leads to a new limit condition in terms of Z where, for reasons that will become clear, we set $\kappa_2 = 0$,

$$T \equiv \frac{\kappa_1 Z}{\kappa_3 Z + \kappa_4} > H_0 \qquad (4.6)$$

We characterize T as a 'control temperature'. Again, H_0, the rate at which the system generates its own topological information, and represents ongoing business as usual. $H(Z)$ is the 'control signal' information rate that must punch through for notice.

For appropriate values of the κ, the relation between T and Z is shown in Fig. 4.3, expected to be monotonic increasing in Z from zero.

The horizontal line in Fig. 4.3 is the DRT limit H_0, the rate at which the system generates topological information by 'doing its own business'. If $T(Z) < H_0$, the signal Z will not have direct effect on behavior. That is, there will be no accession to the rapid mechanisms of consciousness.

We can explore the punctuated accession to consciousness for this model in the presence of 'distraction', as in the previous chapter. Typically, Z itself will be time dependent, rising from zero to some final value according to some adaptation function, which we again take as $dZ/dt = \beta - \alpha Z(t)$, $Z \to \beta/\alpha$. What happens to the dynamics of Fig. 4.2 in a 'distraction' environment? Then, as above, dynamics are approximated by the stochastic differential equation (Protter 2005)

$$dZ_t = (\beta - \alpha Z_t)dt + \sigma Z_t dB_t \qquad (4.7)$$

where, as above, the second term represents the volatility of the system, dB_t representing Brownian noise.

Our central interest, however, is not in Z but in the stochastic behavior of $T(Z) = \kappa_1 Z/(\kappa_3 Z + \kappa_4)$, and this can be done using the Ito Chain Rule (Protter 2005). This

Fig. 4.4 The dark line is the solution equivalence class $\{Z, \ \sigma\}$ and the lighter line the value of Z corresponding the Data Rate Theorem limit H_0 from Fig. 4.3. For $\sigma = 0$, Z is at the nonequilibrium steady state β/α. Rising σ drives the effective value of signal below punctuated detectability

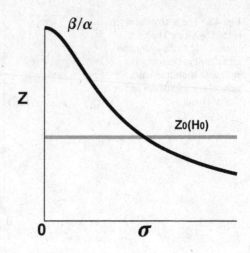

produces a general relation at nonequilibrium steady state as

$$(-Z\alpha + \beta)\left(\frac{\kappa_1}{Z\kappa_3 + \kappa_4} - \frac{\kappa_1 Z\kappa_3}{(Z\kappa_3 + \kappa_4)^2}\right)$$
$$+\frac{\sigma^2 Z^2}{2}\left(-2\frac{\kappa_1 \kappa_3}{(Z\kappa_3 + \kappa_4)^2} + 2\frac{\kappa_1 Z\kappa_3^2}{(Z\kappa_3 + \kappa_4)^3}\right) = 0 \qquad (4.8)$$

The σ-term represents the Ito correction factor indexing the influence of 'distraction' noise.

A typical solution to Eq. (4.8) is shown in Fig. 4.4. This represents a solution equivalence class $\{Z, \ \sigma\}$, along with the value of Z corresponding to the DRT limit H_0 from Fig. 4.3. If $\sigma = 0$, then Z is at the nonequilibrium steady state β/α. As σ increases, the effective value of Z declines until it falls below the punctuated signal entry value. The appearance of an equivalence class again suggests groupoid symmetry breaking at a critical input strength.

4.4   Desensitization/Coma

The identification of $g(Z)$ from Eqs. (2.6), (2.10), (2.13) et seq. as a temperature analog defined via Lambert W-functions suggests possible modification of Eqs. (4.4)–(4.6) through an iteration in $g(Z)$ rather than directly in terms of Z itself, so that

$$H(g) > h(g(Z))H_0$$
$$H(g) \approx \kappa_1 g(Z) + \kappa_2$$
$$T_g \equiv \frac{\kappa_1 g(Z)}{\kappa_3 g(Z) + \kappa_4} \qquad (4.9)$$

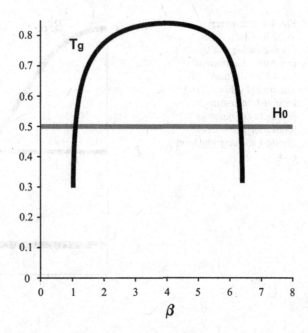

Fig. 4.5 A Yerkes–Dodson inverted-U for punctuated accession to consciousness. Too great a level of arousal causes desensitization

Again assuming an exponential model for the adaptation function, and using Eqs. (2.9) and (2.10), we set $\alpha = 1, C_1 = -3, C_2 = 5, \kappa_j = 1$, and plot another inverted-U arousal graph for T_g versus β in Fig. 4.5. Note here that, if $T_g < H_0$, accession to consciousness fails: at sufficient β, the system becomes desensitized to arousal.

It is also possible to conduct an Ito Chain Rule calculation, finding conditions for which $< dT_g >= 0$. This is given in Fig. 4.6, again taking $dZ/dt = f(Z) = \beta - \alpha Z(t)$ at the peak of Fig. 4.5, $\beta = 4$. For this particular model realization, the system fails at very low levels of σ, compared to the expected failure of the SDE model $dZ_t = f(Z_t)dt + \sigma Z_t dB_t$ at $\sigma = \sqrt{2}$. Again, the critical value $Z_0(H_0)$ is indicated.

4.5 Shadows: Inattentional Blindness

Focusing on T or T_g of the previous section, entry into consciousness requires T-values greater than the Data Rate Theorem limit H_0, the rate at which an embedding environment is generating its own topological information. Typically, the organism must operate under constraints of time t and a resource rate Z that will always be delivered with some delay. Lagrangian optimization under these conditions introduces environmental 'shadow prices', in addition to the burden of delay.

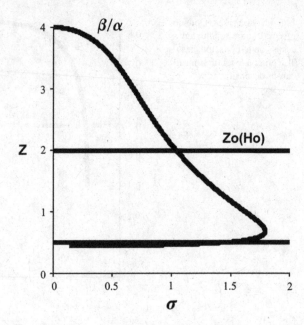

Fig. 4.6 Numerical solution to the Ito Chain Rule calculation for $< dT_g > = 0$ under the conditions of Fig. 4.5 for the peak $\beta = 4$. The critical value $Z_0(H_0)$ is indicated. Increasing probability of unstable bifurcation becomes manifest at a very low level of σ

We suppose the organism is composed of multiple, interacting cognitive sub-components and must optimize a T-value from above across the systems, subject to resource constraints. In addition, we impose inherent delay on the delivery of resources via an exponential relation, so that

$$dZ_k/dt_k = \beta_k - \alpha_k Z_k(t_k) \qquad (4.10)$$

noting that $\lim_{t \to \infty} Z_k(t) = \beta_k/\alpha_k$.

The appropriate Lagrangian schema is then

$$\mathscr{L} = T\left(Z_1(t_1), \ldots, Z_n(t_n)\right) + \lambda\left(t - \sum_k t_k\right) + \mu\left(Z - \sum_k Z_k\right)$$

$$\partial \mathscr{L}/\partial Z_k = \partial T/\partial Z_k - \mu = 0$$

$$\partial \mathscr{L}/\partial t_k = (\partial T/\partial Z_k)dZ_k/dt_k - \lambda = \mu dZ_k/dt_k - \lambda = 0 \quad (4.11)$$

The final expression in Eq. (4.11) gives

$$\frac{\lambda}{\mu} = \beta_k - \alpha_k Z_k(t_k)$$

$$Z_k(t_k) = \frac{\beta_k}{\alpha_k} - \frac{1}{\alpha_k}\frac{\lambda}{\mu} \qquad (4.12)$$

The 'undetermined multipliers' λ and μ, in economic terms, are 'shadow prices' imposed by environmental externalities (e.g., Jin et al. 2008). If enough α_k are small, and the shadow price ratio is large, then some of the Z_k will be below critical values, and the organism will likely be blind to environmental change, e.g., as inattentional blindness, and so on.

The result even holds if there is an imposed exponentially-distributed delay, i.e., if

$$dZ/dt = \beta - \alpha \int_0^t Z(t-\tau)m\exp[-m\tau]d\tau \qquad (4.13)$$

and something similar can be shown under fixed delays as well, that is, replacing t_k with $t_k - \delta_k$ in Eq. (4.10).

References

Cohen I (2000) Tending Adam's garden: evolving the cognitive immune self. Academic, New York

Dehaene S, Changeux J (2011) Experimental and theoretical approaches to conscious processing. Neuron 70:200–227

English T (1996) Evaluation of evolutionary and genetic optimizers: no free lunch. In: Fogel L, Angeline P, Back T (eds) Evolutionary programming V: proceedings of the fifth annual conference on evolutionary programming. MIT Press, Cambridge, pp 163–169

Glazebrook JF, Wallace R (2009) Rate distortion manifolds as model spaces for cognitive information. Informatica 33:309–345

Hahn P (1978) The regular representations of measure groupoids. Trans Am Math Soc 242:35–53

Jin H, Hu Z, Zhou X (2008) A convex stochastic optimization problem arising from portfolio selection. Math Financ 18:171–183

Mackey GW (1963) Ergodic theory, group theory, and differential geometry. Proc Natl Acad Sci USA 50:1184–1191

Newman M (2010) Networks: an introduction. Oxford University Press, New York

Oeckl R (2003) Renormalization of discrete models without background. Nucl Phys B 657:107–138

Protter P (2005) Stochastic integration and differential equations: a new approach, 2nd edn. Springer, New York

Series C (1977) Ergodic actions of product groups. Pac J Math 70:519–534

Spenser J (2010) The giant component: a golden anniversary. Not Am Math Soc 57:720–724

Tateishi A, Hanel R, Thurner S (2013) The transformation groupoid structure of the q-Gaussian family. Phys Lett A 377:1804–1809

Wallace R (2005) Consciousness: a mathematical treatment of the global neuronal workspace model. Springer, New York

Wallace R (2012) Consciousness, crosstalk, and the mereological fallacy: an evolutionary perspective. Phys Life Rev 9:426–453

Wallace R (2018) New statistical models of nonergodic cognitive systems and their pathologies. J Theor Biol 436:72–78

Wallace R (2020) Cognitive dynamics on Clausewitz landscapes: the control and directed evolution of organized conflict. Springer, New York

Wallace R (2021a) Toward a formal theory of embodied cognition. Biosystems. https://doi.org/10.1016/j.biosystems.2021.104356

Wallace R (2021b) Military scientism and its discontents. JDMS. https://doi.org/10.1177/15485129211014281

Wallace R, Wallace D (2016) Gene expression and its discontents: the social production of chronic disease, 2nd edn. Springer, New York

Wilson K (1971) Renormalization group and critical phenomena. I. Renormalization group and the Kadanoff scaling picture. Phys Rev B 4:3174–3183

Wolpert D, MacReady W (1997) No free lunch theorems for optimization. IEEE Trans Evol Comput 1:67–82

Chapter 5
Discussion and Conclusions

As a remark made above implies—that for information dynamics there is no microreversibility, and hence no 'Onsager Reciprocal Relations'—cognitive phenomena are likely to be far different from physical processes, although undoubtedly constrained by them. The 'prime groupoid phase transition' is both 'obvious' and unexpected. The 'Kadanoff Picture' of phase transition in cognition is similarly plagued with 'biological renormalizations' that may, in fact, be tunable (Wallace 2005). Underlying this is the matter of 'fundamental symmetries' and 'symmetry-breaking' in cognition. These 'symmetries' will, particularly for the nonergodic cognitive phenomena likely to dominate real-world dynamics, involve equivalence classes of dynamic paths, the long $x_j = \{x_1^j, ..., x_n^j, ...\}$ discussed above. Equivalence class properties can be expressed in terms of groupoids, essentially groups for which there is not necessarily a product defined between all element pairs. The symmetry-breaking of phase transitions familiar from physical theory then becomes a matter of transitions between groupoid structures, at least for stationary sources and adiabatic approximations by stationary sources.

That is, symmetry-breaking in cognition should be considered as fundamental to the study of cognitive process—including consciousness—as it is to physical theory. The symmetries are, however, much different than those familiar from physical theory. One may, perhaps in lifting the requirement that the systems be stationary, be driven to even more general symmetry structures than groupoids, for example, small categories and semigroupoids, in the context of dynamics characterized in terms of formal algebraic power series. This work remains to be done.

To reiterate, 'except on a set of measure zero' implies some primordial symmetry breaking.

There is support for this perspective in the literature. Recall the 'information theory chain rule' from Eq. (2.2) (Cover and Thomas 2006). For two stationary, ergodic information sources X_1 and X_2, the joint uncertainty must be less than the

R. Wallace, *Consciousness, Cognition and Crosstalk: The Evolutionary Exaptation of Nonergodic Groupoid Symmetry-Breaking*, SpringerBriefs in Computational Intelligence, https://doi.org/10.1007/978-3-030-87219-9_5

sum of their independent uncertainties:

$$H(X_1) + H(X_2) \geq H(X_1, X_2) \tag{5.1}$$

Let G be any finite group and G_1, G_2 be subgroups of G. Take $|G|$ to be the order of the group, i.e., the number of elements. The intersection $G_1 \cap G_2$ is also a subgroup and a group inequality can be derived analogous to Eq. (5.1):

$$\log\left(\frac{|G|}{|G_1|}\right) + \log\left(\frac{|G|}{|G_2|}\right) \geq \log\left(\frac{|G|}{|G_1 \cap G_2|}\right) \tag{5.2}$$

Yeung (2008) assigns a probability via a pseudorandom variate related to a group G as $\Pr[X = a] = 1/|G|$, allowing construction of a group-characterized information source. Yeung (2008) establishes a one-to-one correspondence between unconstrained information inequalities, extensions of Eq. (5.1), and finite group inequalities. That is, unconstrained inequalities can be proved by techniques in group theory, and many group-theoretic inequalities can be proven by methods of information theory.

We suggest here, in a similar manner, that nonergodic information sources and their dynamics are intimately associated with groupoid algebras. Less regular information processes may require even more general algebraic structures. Some further thoughts in this direction are presented in the Mathematical Appendix.

We have, then, outlined a mathematical treatment of embodied consciousness—really, the only kind evolution can give us—that, while abducting (ultimately, non-linear) nonequilibrium thermodynamics and Kadanoff theory, remains true to the asymptotic limit theorems of information and control theories. The underlying example for this is the abduction of classical mechanics into both quantum theory and general relativity, albeit in markedly different directions. The observation of Feynman (2000) and many others that information is a form of free energy permits these abductions, in the context of new, iterative, Morse Theory free energy and entropy analogs. Application of these methods to psychopathologies can be found in Wallace (2017), and to failure of artificial intelligence under stress in Wallace (2020a).

This work differs significantly from the earlier analyses by Wallace (2005, 2012), who studied similar dynamics, but focused on adiabatically piecewise stationary ergodic (APSE) information sources dual to cognitive processes. Here, an iterated free energy Morse Function is defined through Eq. (2.4) for nonergodic systems, permitting greater latitude in modeling dynamic behavior, still however, in the context of an adiabatic approximation that sees dual information sources as 'sufficiently' stationary with respect to much slower embedding dynamics.

This iterated 'free energy' approach differs from Friston's free energy formalism (e.g., Bogacz 2017) by avoiding a fundamental contradiction, i.e., not invoking minimization of free energy measures for neural systems that actually require rates of metabolic free energy supply an order of magnitude greater than for other kinds of tissue. The argument here that most parallels Friston's regards efficiency of cognition, as in Fig. 3.3, suggesting an evolutionary necessity for highly parallel address of difficult cognitive problems.

Further, the underlying perspective of this line of research differs from Integrated Information Theory (IIT) by actually hewing closely to the asymptotic limit theorems of both control and information theories, and by explicit recognition that consciousness, like immune function and insect wings, is an evolutionary adaptation specific to organisms, and not a general property to be assigned across physical systems. While it may be possible to construct computing machines having any number of rapidly-tunable neural global workspace analogs, there is no panpsychic aether.

Indeed, consciousness in higher animals appears as a necessarily stripped-down, greatly simplified, high-speed example of much slower, but far more general, processes—like immune function, wound healing, and gene expression/regulation—that entertain multiple, simultaneous tunable spotlight 'global workspaces' (Wallace 2012). All such have emerged through evolutionary exaptations of the inevitable crosstalk afflicting information processes through a kind of 'second law' leakage necessarily associated with information as a form of free energy.

Such an evolutionary perspective represents a fundamental reorientation in consciousness studies, stripping the subject of various deep, culturally-driven—but scientifically irrelevant—social constructs.

In sum, without identifying consciousness as a weird, new, form of matter, without mind/body dualism, without the *ignis fatuus* of the 'hard problem', 'qualia', and like conceits, the asymptotic limit theorems of information and control theories permit construction of models recognizably similar to the empirical pictures Bernard Baars and others have drawn of high level mental phenomena. That being said, we are constrained by the warning of the mathematical ecologist E.C. Pielou (1977), that the purpose of mathematical models is new speculation, not new knowledge, which can only arise from observation and experiment.

Most particularly, then, the probability models outlined here should be seen as analogs to such 'models' as the Central Limit and associated asymptotic theorems that are the foundations of statistical tools including t-tests, regression equations, and so on. Such tools, among other uses, provide important benchmarks against which to compare experimental and observational results, and new knowledge is as likely to come from their failures as from their successes.

Following Wallace (2017, Sect. 7.7), we speculate further that such tools might well aid in the understanding and treatment of the many pathologies afflicting cognitive process at and across the various scales and levels of organization that characterize the living state.

References

Bogacz R (2017) A tutorial on the free-energy framework for modelling perception and learning. J Math Psychol 76:198–211

Cover T, Thomas J (2006) Elements of information theory, 2nd edn. Wiley, New York

Feynman R (2000) Lectures on computation. Westview Press, New York

Pielou E (1977) Mathematical ecology. Wiley, New York

Wallace R (2005) Consciousness: a mathematical treatment of the global neuronal workspace model. Springer, New York

Wallace R (2012) Consciousness, crosstalk, and the mereological fallacy: an evolutionary perspective. Phys Life Rev 9:426–453

Wallace R (2017) Computational psychiatry: a systems biology approach to the epigenetics of mental disorders. Springer, New York

Wallace R (2020a) How AI founders on adversarial landscapes of fog and friction. J Def Model Simul. https://doi.org/10.1177/1548512920962227

Yeung H (2008) Information theory and network coding. Springer, New York

Chapter 6
Mathematical Appendix

6.1 Groupoids

We following Brown (1992). Consider a directed line segment in one component, written as the source on the left and the target on the right.

Two such arrows can be composed to give a product **ab** if and only if the target of **a** is the same as the source of **b**

Brown puts it this way,

> One imposes the geometrically obvious notions of associativity, left and right identities, and inverses. Thus a groupoid is often thought of as a group with many identities, and the reason why this is possible is that the product **ab** is not always defined.
>
> We now know that this apparently anodyne relaxation of the rules has profound consequences... [since] the algebraic structure of product is here linked to a geometric structure, namely that of arrows with source and target, which mathematicians call a *directed graph*.

Cayron (2006) elaborates this:

> A group defines a structure of actions without explicitly presenting the objects on which these actions are applied. Indeed, the actions of the group G applied to the identity element e implicitly define the objects of the set G by ge = g; in other terms, in a group, actions and objects are two isomorphic entities. A groupoid enlarges the notion of group by explicitly introducing, in addition to the actions, the objects on which the actions are applied. By this approach, many identities may exist (they correspond to the actions that leave an object invariant).

R. Wallace, *Consciousness, Cognition and Crosstalk: The Evolutionary Exaptation of Nonergodic Groupoid Symmetry-Breaking*, SpringerBriefs in Computational Intelligence, https://doi.org/10.1007/978-3-030-87219-9_6

Stewart (2017) describes something of the underlying mechanics by which symmetry changes in general may be expressed:

> Spontaneous symmetry-breaking is a common mechanism for pattern formation in many areas of science. It occurs in a symmetric dynamical system when a solution of the equations has a smaller symmetry group than the equations themselves... This typically happens when a fully symmetric solution becomes unstable and branches with less symmetry bifurcate.

It is of particular importance that equivalence class decompositions permit construction of groupoids in a highly natural manner.

Weinstein (1996) and Golubitsky and Stewart (2006) provide more details on groupoids and on the relation between groupoids and bifurcations.

An essential point is that, since there are no necessary products between groupoid elements, 'orbits', in the usual sense, disjointly partition groupoids into 'transitive' subcomponents.

Application of groupoid formalism to the homochirality problem, with a more detailed overview of 'biological' groupoid algebra, can be found in Wallace (2011).

6.2   The Data Rate Theorem

Real-world environments are inherently unstable. Organisms (and organizations), to survive, must exert a considerable measure of control over them. These control efforts range from immediate responses to changing patterns of threat and affordance, through niche construction, and, in higher animals, elaborate, highly persistent, social and sociocultural structures. Such necessity of control can, in some measure, be represented by a powerful asymptotic limit theorem of probability theory different from, but as fundamental as, the Central Limit Theorem. It is called the Data Rate Theorem, first derived as an extension of the Bode Integral Theorem of signal theory.

Consider a reduced model of a control system as follows:

For the inherently unstable system of Fig. 6.1, assume an initial n-dimensional vector of system parameters at time t, as x_t. The system state at time $t + 1$ is then—near a presumed nonequilibrium steady state—determined by the first-order relation

$$x_{t+1} = \mathbf{A}x_t + \mathbf{B}u_t + W_t \tag{6.1}$$

In this approximation, $\mathbf{A}$ and $\mathbf{B}$ are taken as fixed n-dimensional square matrices. u_t is a vector of control information, and W_t is an n-dimensional vector of Brownian white noise.

According to the DRT, if H is a rate of control information sufficient to stabilize an inherently unstable control system, then it must be greater than a minimum, H_0,

$$H > H_0 \equiv \log[|\det[\mathbf{A}^m]|] \tag{6.2}$$

Fig. 6.1 The reduced model
of an inherently unstable
system stabilized by a
control signal U_t

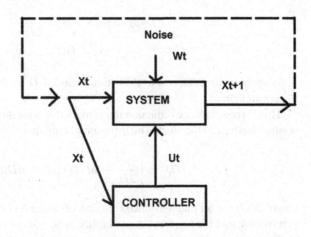

where det is the determinant of the subcomponent $\mathbf{A}^m$—with $m \leq n$—of the matrix
$\mathbf{A}$ having eigenvalues ≥ 1. H_0 is defined as the rate at which the unstable system
generates 'topological information' on its own.

If this inequality is violated, stability fails.

A somewhat different derivation of the DRT, via the 'convexity' inherent to the
Rate Distortion function, is also possible for stationary ergodic systems, as follows.

Assume control information is supplied to an inherently unstable system at a rate
H. Let $R(D)$ be a classic Rate Distortion Function inferred from Fig. 6.1, seen as
characterizing the relation between system intent and system operational effect. A
distortion measure D is defined as a scalar measure of the disjunction between intent
and impact of the regulator. We take R_t as the Rate Distortion Function at time t,
imposing conditions of noise and volatility so that system dynamics are described
by the stochastic differential equation

$$dR_t = f(R_t, t)dt + bR_t dB_t \qquad (6.3)$$

where dB_t represents Brownian noise, f is an appropriate function, and b a parameter.

We take $H(R_t, t)$ to be the incoming rate of control information needed to impose
control, and apply the Black–Scholes argument (Wallace 2020c, Sect. 14.4), expanding H in terms of R by using the Ito Chain Rule on Eq. (6.3). At nonequilibrium
steady state, 'it is not difficult to show' that

$$H_{nss} \approx \kappa_1 R + \kappa_2 \qquad (6.4)$$

for appropriate constants κ_i.

For a Gaussian channel, $R = 1/2 \log[\sigma^2/D]$, and, recalling Feynman's (2000)
characterization of information as a form of free energy, we can define an entropy
as the Legendre transform of R, i.e., $S = -R(D) + DdR(D)/dD$. This leads to the
usual Onsager approximation for the dynamics of the distortion scalar D:

$$dD/dt \approx \mu dS/dD = \frac{\mu}{2D(t)}$$

$$D(t) \approx \sqrt{\mu t} \qquad (6.5)$$

Without control, the scalar distortion measure D undergoes a familiar diffusion to catastrophe.

This correspondence reduction to ordinary diffusion suggests an iterative approximation leading to the stochastic differential equation

$$dD_t = \left[\frac{\mu}{2D_t} - M(H) \right] dt + \beta D_t dB_t \qquad (6.6)$$

where $M(H)$ is an undetermined function of the rate at which control information is provided, and the last term is a volatility in Brownian noise.

The nonequilibrium steady state expectation of Eq. (6.4) is simply

$$D_{nss} = \frac{\mu}{2M(H)} \qquad (6.7)$$

Application of the Ito Chain Rule to D_t^2 via Eq. (6.5) implies the necessary condition for stability in variance is

$$M(H) \geq \beta \sqrt{\mu} \qquad (6.8)$$

From Eqs. (6.4) and (6.7), however,

$$M(H) \geq \frac{\mu}{2\sigma^2} \exp[2(H - \kappa_2)/\kappa_1] \geq \beta \sqrt{\mu} \qquad (6.9)$$

giving an explicit necessary condition for stability in second order as

$$H \geq \frac{\kappa_1}{2} \log \left(\frac{2\beta\sigma^2}{\sqrt{\mu}} \right) + \kappa_2 \equiv H_0 \qquad (6.10)$$

Other channel configurations with different explicit algebraic expressions for the RDF will nonetheless have similar results as a consequence of the inherent convexity of the RDF.

6.3 Morse Theory

Morse theory examines relations between analytic behavior of a function—the location and character of its critical points—and the underlying topology of the manifold on which the function is defined. We are interested in a number of such functions, for example a 'free energy' constructed from information source uncertainties on a parameter space and 'second order' iterations involving parameter manifolds determining critical behavior. These can be reformulated from a Morse theory perspective. Here we follow closely Pettini (2007).

The essential idea of Morse theory is to examine an n-dimensional manifold M as decomposed into level sets of some function $f : M \rightarrow \mathbf{R}$ where $\mathbf{R}$ is the set of real numbers. The a-level set of f is defined as

$$f^{-1}(a) = \{x \in M : f(x) = a\},$$

the set of all points in M with $f(x) = a$. If M is compact, then the whole manifold can be decomposed into such slices in a canonical fashion between two limits, defined by the minimum and maximum of f on M. Let the part of M below a be defined as

$$M_a = f^{-1}(-\infty, a] = \{x \in M : f(x) \leq a\}.$$

These sets describe the whole manifold as a varies between the minimum and maximum of f.

Morse functions are defined as a particular set of smooth functions $f : M \rightarrow \mathbf{R}$ as follows. Suppose a function f has a critical point x_c, so that the derivative $df(x_c) = 0$, with critical value $f(x_c)$. Then f is a Morse function if its critical points are nondegenerate in the sense that the Hessian matrix of second derivatives at x_c, whose elements, in terms of local coordinates are

$$H_{i,j} = \partial^2 f / \partial x^i \partial x^j,$$

has rank n, which means that it has only nonzero eigenvalues, so that there are no lines or surfaces of critical points and, ultimately, critical points are isolated.

The index of the critical point is the number of negative eigenvalues of H at x_c.

A level set $f^{-1}(a)$ of f is called a critical level if a is a critical value of f, that is, if there is at least one critical point $x_c \in f^{-1}(a)$.

Again following Pettini (2007), the essential results of Morse theory are:

1. If an interval $[a, b]$ contains no critical values of f, then the topology of $f^{-1}[a, v]$ does not change for any $v \in (a, b]$. Importantly, the result is valid even if f is not a Morse function, but only a smooth function.

2. If the interval $[a, b]$ contains critical values, the topology of $f^{-1}[a, v]$ changes in a manner determined by the properties of the matrix H at the critical points.

3. If $f : M \rightarrow \mathbf{R}$ is a Morse function, the set of all the critical points of f is a discrete subset of M, i.e. critical points are isolated. This is Sard's Theorem.

4. If $f : M \rightarrow \mathbf{R}$ is a Morse function, with M compact, then on a finite interval $[a, b] \subset \mathbf{R}$, there is only a finite number of critical points p of f such that $f(p) \in [a, b]$. The set of critical values of f is a discrete set of $\mathbf{R}$.

5. For any differentiable manifold M, the set of Morse functions on M is an open dense set in the set of real functions of M of differentiability class r for $0 \leq r \leq \infty$.

6. Some topological invariants of M, that is, quantities that are the same for all the manifolds that have the same topology as M, can be estimated and sometimes

computed exactly once all the critical points of f are known: Let the Morse numbers $\mu_i (i = 1, \ldots, m)$ of a function f on M be the number of critical points of f of index i, (the number of negative eigenvalues of H). The Euler characteristic of the complicated manifold M can be expressed as the alternating sum of the Morse numbers of any Morse function on M,

$$\chi = \sum_{i=0}^{m} (-1)^i \mu_i.$$

The Euler characteristic reduces, in the case of a simple polyhedron, to

$$\chi = V - E + F$$

where V, E, and F are the numbers of vertices, edges, and faces in the polyhedron.

7. Another important theorem states that, if the interval $[a, b]$ contains a critical value of f with a single critical point x_c, then the topology of the set M_b defined above differs from that of M_a in a way which is determined by the index, i, of the critical point. Then M_b is homeomorphic to the manifold obtained from attaching to M_a an i-handle, i.e., the direct product of an i-disk and an $(m - i)$-disk.

Again, Pettini (2007) contains both mathematical details and further references. See, for example, Matsumoto (1997).

6.4   Higher Dimensional Resource Systems

We assumed that resource delivery is sufficiently characterized by a single scalar parameter Z, mixing material resource/energy supply with internal and external flows of information. Real world conditions will likely be far more complicated. Invoking a perspective analogous to Principal Component Analysis, there may be several independent pure or composite entities irreducibly driving system dynamics. It may then be necessary to replace the scalar Z by an n-dimensional vector $\mathbf{Z}$ having orthogonal components that together account for much of the total variance—in a sense—of the rate of supply of essential resources. The dynamic equations (6.9) (and Eq. (6.11)) must then be represented in vector form:

$$F(\mathbf{Z}) = -\log\left(h(g(\mathbf{Z}))\right) g(\mathbf{Z})$$
$$S = -F + \mathbf{Z} \cdot \nabla_{\mathbf{Z}} F$$
$$\partial \mathbf{Z}/\partial t \approx \hat{\mu} \cdot \nabla_{\mathbf{Z}} S = f(\mathbf{Z})$$
$$-\nabla_{\mathbf{Z}} F + \nabla_{\mathbf{Z}}(\mathbf{Z} \cdot \nabla_{\mathbf{Z}} F) =$$
$$\hat{\mu}^{-1} \cdot f(\mathbf{Z}) \equiv f^*(\mathbf{Z})$$
$$\left((\partial^2 F/\partial z_i \partial z_j)\right) \cdot \mathbf{Z} = f^*(\mathbf{Z})$$
$$\left((\partial^2 F/\partial z_i \partial z_j)\right)|_{\mathbf{Z}_{nss}} \cdot \mathbf{Z}_{\mathbf{nss}} = 0 \qquad\qquad (6.11)$$

F, g, h, and S are, again, scalar functions, but $\hat{\mu}$ is an n-dimensional square matrix of diffusion coefficients. The matrix $((\partial F / \partial z_i \partial z_j))$ is the obvious n-dimensional square matrix of second partial derivatives, and $f(\mathbf{Z})$ is a vector function. The last relation imposes a nonequilibrium steady state condition, i.e. $f^*(\mathbf{Z}_{nss}) = \mathbf{0}$.

For a detailed two-dimensional example—requiring Lie group symmetry methods—see Wallace (2020a, Sect. 7; 2020b).

6.5 Distraction and Iterated Free Energy

Here, rather than the cognition rate $L(Z)$ that leads to intractable equations, we apply the Ito Chain Rule to the iterated free energy analog F defined by the relations of Eq. (2.10), taking $f(Z) = \beta - \alpha Z$. Then

$$F(Z) = \ln(Z) Z\beta - Z\beta - \frac{Z^2 \alpha}{2} + C_1 Z + C_2 \tag{6.12}$$

Application of the Ito Chain Rule to $F(Z)$ gives, after some calculation, an expression for $< dF_t >= 0$ as a nonequilibrium steady state relation in Z

$$(-\alpha Z + \beta)(\beta \ln(Z) - \alpha Z + C_1) + \frac{\sigma^2 Z^2}{2}\left(\frac{\beta}{Z} - \alpha\right) = 0 \tag{6.13}$$

Equation (6.13) has the exact solutions

$$Z = \beta / \alpha$$

$$Z = -2\frac{\beta}{-\sigma^2 + 2\alpha} W\left(n, 1/2\frac{\sigma^2 - 2\alpha}{\beta}e^{-\frac{C_1}{\beta}}\right) \tag{6.14}$$

where, again, $W(n, x)$ is the Lambert W-function of orders 0 or -1.

This result is shown in Fig. 6.2, taking $\alpha = 1$, $\beta = 3$ $C_1 = -1$, $C_2 = 1$.

Again, the system explodes as $\sigma^2/2 \to \alpha$, but now the Lambert W-function imposes a real-value condition at $-\exp[-1]$ so that a bifurcation instability sets in if

$$\sigma > \sqrt{2}\sqrt{e^{-\frac{C_1}{\beta}}\left(e^{-\frac{C_1}{\beta}}\alpha - e^{-1}\beta\right)\left(e^{-\frac{C_1}{\beta}}\right)^{-1}} \tag{6.15}$$

For the conditions of Fig. 6.2, that triggering noise level is far below the limit $\sigma^2/2 \to \alpha$, expected from the exponential relation $dZ/dt = f(Z(t)) = \beta - \alpha Z(t)$. This transition is closely analogous to the ways in which 'noise' can drive phase transitions in simple physical systems, studied by Van den Broeck et al. (1994, 1997), Horsthemeke and Lefever (2006), Wallace (2016), and many others.

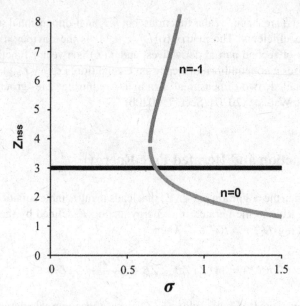

Fig. 6.2 Phase transition diagram for the iterated free energy function F under the exponential adaptation relation $\partial Z/\partial t = f(Z) = \beta - \alpha Z$. Here, $\alpha = 1$, $\beta = 3$, $C_1 = -1$, $C_2 = 1$. The horizontal line is the solution $Z = \beta/\alpha$. The upper trace corresponds to the $n = -1$ branch of the Lambert W-function, and the lower to the $n = 0$ branch. While Eq. (6.14) again explodes in variance for $\sigma^2/2 \geq \alpha = 1$, $\sigma \geq \sqrt{2} = 1.414\ldots$, in this particular model realization, $F(Z)$ undergoes a bifurcation instability at the considerably lower value $\sigma > 0.646851\ldots$

6.6   Biological Renormalizations

Here, we adapt the renormalization scheme of Wallace (2005), focused on a stationary, ergodic, information source H, to the generalized free energy associated with nonergodic cognition.

Equation (4.2) states that the information source and the correlation length, the degree of coherence on the underlying network, scale under renormalization clustering in chunks of size R as

$$F[\omega(R), J(R)] = \mathscr{F}(R)F[\omega(1), J(1)]$$

$$\chi[\omega(R), J(R)]R = \chi[\omega(1), J(1)]$$

with $F(1) = 1$.

Differentiating these two equations with respect to R, so that the right hand sides are zero, and solving for $d\omega(R)/dR$ and $dJ(R)/dR$ gives, after some manipulation,

$$dw_R/dR = u_1 d \log(\mathscr{F})/dR + u_2/R$$
$$dJ_R/dR = v_1 J_R d \log(\mathscr{F})/dR + \frac{v_2}{R} J_R \qquad (6.16)$$

The $u_i, v_i, i = 1, 2$ are functions of $\omega(R), J(R)$, but not explicitly of R itself.

We expand these equations about the critical value $\omega_R = \omega_C$ and about $J_R = 0$, obtaining

$$dw_R/dR = (\omega_R - \omega_C) y d \log(\mathscr{F})/dR + (\omega_R - \omega_C) z/R$$
$$dJ_R/dR = w J_R d \log(\mathscr{F})/dR + x J_R/R \qquad (6.17)$$

The terms $y = du_1/dw_R|_{\omega_R = \omega_C}, z = du_2/dw_R|_{\omega_R = \omega_C}, w = v_1(\omega_C, 0), x = v_2$ $(\omega_C, 0)$ are constants.

Solving the first of these equations gives

$$\omega_R = \omega_C + (\omega - \omega_C) R^z \mathscr{F}(R)^y \qquad (6.18)$$

again remembering that $\omega_1 = \omega, J_1 = J, \mathscr{F}(1) = 1$.

Wilson's (1971) essential trick is to iterate on this relation, which is supposed to converge rapidly near the critical point, assuming that for ω_R near ω_C, we have

$$\omega_C/2 \approx \omega_C + (\omega - \omega_C) R^z \mathscr{F}(R)^y \qquad (6.19)$$

We iterate in two steps, first solving this for $\mathscr{F}(R)$ in terms of known values, and then solving for R, finding a value R_C that we then substitute into the first of Eqs. (4.2) to obtain an expression for $F[\omega, 0]$ in terms of known functions and parameter values.

The first step gives the general result

$$\mathscr{F}(R_C) \approx \frac{[\omega_C/(\omega_C - \omega)]^{1/y}}{2^{1/y} R_C^{z/y}} \qquad (6.20)$$

Solving this for R_C and substituting into the first expression of Eq. (6.31) gives, as a first iteration of a far more general procedure (Shirkov and Kovalev 2001), the result

$$F[\omega, 0] \approx \frac{F[\omega_C/2, 0]}{\mathscr{F}(R_C)} = \frac{F_0}{\mathscr{F}(R_C)}$$
$$\chi(\omega, 0) \approx \chi(\omega_C/2, 0) R_C = \chi_0 R_C \qquad (6.21)$$

giving the essential relationships.

Note that a power law of the form $\mathscr{F}(R) = R^m, m = 3$, which is the direct physical analog, may not be biologically reasonable, since it says that 'language richness', in a general sense, can grow very rapidly as a function of increased network size. Such rapid growth is simply not observed in cognitive process.

Taking the biologically realistic example of non-integral 'fractal' exponential growth,

$$\mathscr{F}(R) = R^\delta \tag{6.22}$$

where $\delta > 0$ is a real number which may be quite small, equation we can be solve for R_C, obtaining

$$R_C = \frac{[\omega_C/(\omega_C - \omega)]^{[1/(\delta y + z)]}}{2^{1/(\delta y + z)}} \tag{6.23}$$

for ω near ω_C. Note that, for a given value of y, one might characterize the relation $\alpha \equiv \delta y + z = \text{constant}$ as a 'tunable universality class relation' in the sense of Albert and Barabasi (2002).

Substituting this value for R_C back gives a complex expression for F, having three parameters: δ, y, z.

A more biologically interesting choice for $\mathscr{F}(R)$ is a logarithmic curve that 'tops out', for example

$$\mathscr{F}(R) = m \log(R) + 1 \tag{6.24}$$

Again $\mathscr{F}(1) = 1$.

Using a computer algebra program to solve for R_C gives

$$R_C = \left[\frac{Q}{W[0, Q \exp(z/my)]} \right]^{y/z} \tag{6.25}$$

where

$$Q \equiv (z/my) 2^{-1/y} [\omega_C/(\omega_C - \omega)]^{1/y}$$

Again, W(n, x) is the Lambert W-function of order n.

An asymptotic relation for $\mathscr{F}(R)$ would be of particular biological interest, implying that 'language richness' increases to a limiting value with population growth. Taking

$$\mathscr{F}(R) = \exp[m(R - 1)/R] \tag{6.26}$$

gives a system which begins at 1 when $R = 1$, and approaches the asymptotic limit $\exp(m)$ as $R \to \infty$. Computer algebra finds

$$R_C = \frac{my/z}{W[0, A]} \tag{6.27}$$

where

$$A \equiv (my/z) \exp(my/z) [2^{1/y} [\omega_C/(\omega_C - \omega)]^{-1/y}]^{y/z}$$

These developments suggest the possibility of taking the theory significantly beyond arguments by abduction from simple physical models.

6.7 The Tuning Theorem

Here, we closely follow Wallace (2017, Sect. 12.2).

Messages from an information source, taken as a sequence of symbols x_j from some particular alphabet, with each having probabilities P_j associated with a random variable X, are subsequently 'encoded' into the language of a 'transmission channel'. The channel is a random variable Y with symbols y_k, having probabilities P_k. The 'encoding' is possibly with error.

The recipient of the symbol y_k then retranslates it (without error) into some x_k of the original alphabet, which may or may not be the same as the x_j that was sent.

More formally, the message sent along the channel is characterized by a random variable X having the distribution $P(X = x_j) = P_j, j = 1, \ldots, M$.

The channel through which the message is sent is characterized by a second random variable Y having the distribution $P(Y = y_k) = P_k, k = 1, \ldots, \ldots L$.

Let the joint probability distribution of X and Y be defined as $P(X = x_j, Y = y_k) = P(x_j, y_k) = P_{j,k}$ and the conditional probability of Y given X as $P(Y = y_k | X = x_j) = P(y_k | x_j)$.

Then the Shannon uncertainty of X and Y independently, and the joint uncertainty of X and Y together, are defined respectively as

$$H(X) = - \sum_{j=1}^{M} P_j \log(P_j)$$

$$H(Y) = - \sum_{k=1}^{L} P_k \log(P_k)$$

$$H(X, Y) = - \sum_{j=1}^{M} \sum_{k=1}^{L} P_{j,k} \log(P_{j,k}) \tag{6.28}$$

The *conditional uncertainty* of Y given X is then defined as

$$H(Y|X) = - \sum_{j=1}^{M} \sum_{k=1}^{L} P_{j,k} \log[P(y_k | x_j)] \tag{6.29}$$

Recall that, for any two stochastic variates X and Y, $H(Y) \geq H(Y|X)$, since knowledge of X generally gives some knowledge of Y. Equality then occurs only in the case of stochastic independence.

Since $P(x_j, y_k) = P(x_j)P(y_k | x_j)$, then $H(X|Y) = H(X, Y) - H(Y)$.

The information transmitted by translating the variable X into the channel transmission variable Y—possibly with error—and then retranslating without error the transmitted Y back into X is defined as

$$I(X|Y) \equiv H(X) - H(X|Y) = H(X) + H(Y) - H(X, Y) \tag{6.30}$$

See Cover and Thomas (2006) for details. Note that that, if there is no uncertainty in X given the channel Y, then there is no loss of information through transmission. In general this will not be true, and in that lies the centrality of the theory.

Given a fixed vocabulary for the transmitted variable X, and a fixed vocabulary and probability distribution for the channel Y, it is possible to vary the probability distribution of X in such a way as to maximize the information sent. The capacity of the channel is defined as

$$C \equiv \max_{P(X)} I(X|Y) \tag{6.31}$$

subject to the condition $\sum P(X) = 1$.

The central idea of the Shannon Coding Theorem for sending a message with arbitrarily small error along the channel Y at any rate $R < C$ is to encode it in longer and longer 'typical' sequences of the variable X; that is, those sequences whose distribution of symbols approximates the probability distribution $P(X)$ above which maximizes C.

If $S(n)$ is the number of such 'typical' sequences of length n, then $\log[S(n)] \approx nH(X)$ where $H(X)$ is the uncertainty of the stochastic variable defined above. Some consideration shows that $S(n)$ is much less than the total number of possible messages of length n. Thus, as $n \to \infty$, only a vanishingly small fraction of all possible messages is meaningful in this sense. This observation allows the Coding Theorem to work so well. The prescription is to encode messages in typical sequences, which are sent at very nearly the capacity of the channel. As the encoded messages become longer and longer, their maximum possible rate of transmission without error approaches channel capacity as a limit.

This approach is inverted to give a 'tuning theorem' variant of the coding theorem: The channel is made typical with respect to the message.

Telephone lines, optical wave guides and the tenuous plasma through which a planetary probe transmits data to earth may all be viewed in traditional information-theoretic terms as a noisy channel around which we must structure a message so as to attain an optimal error-free transmission rate.

Telephone lines, wave guides and interplanetary plasmas are, relatively speaking, fixed on the timescale of most messages, as are most sociogeographic networks—another 'adiabatic approximation'. Indeed, the capacity of a channel, is defined by varying the probability distribution of the message X so as to maximize $I(X|Y)$.

Suppose the message X is considered so critical that its probability distribution must remain fixed. The new trick is to then fix the distribution $P(x)$ but modify the channel—tune it—so as to maximize $I(X|Y)$. The dual channel capacity C^* can then be defined as

$$C^* \equiv \max_{P(Y),P(Y|X)} I(X|Y) \tag{6.32}$$

But $C^* = \max_{P(Y),P(Y|X)} I(Y|X)$ since $I(X|Y) = H(X) + H(Y) - H(X,Y) = I(Y|X)$.

Thus, in a purely formal mathematical sense, the message transmits the channel, and there will indeed be, according to the Coding Theorem, a channel distribution $P(Y)$ which maximizes C^*.

One may do better than this, however, by modifying the channel matrix $P(Y|X)$. Since $P(y_j) = \sum_{i=1}^{M} P(x_i)P(y_j|x_i)$, then $P(Y)$ is entirely defined by the channel matrix $P(Y|X)$ for fixed $P(X)$ and $C^* = \max_{P(Y),P(Y|X)} I(Y|X) = \max_{P(Y|X)} I(Y|X)$.

Calculating C^* requires maximizing the complicated expression $I(X|Y) = H(X) + H(Y) - H(X, Y)$ which contains products of terms and their logs, subject to constraints that the sums of probabilities are 1 and each probability is itself between 0 and 1. Maximization is done by varying the channel matrix terms $P(y_j|x_i)$ within the constraints. This is a challenging problem in nonlinear optimization. However, for the special case $M = L$, C^* may be found by inspection:

If $M = L$, then choose $P(y_j|x_i) = \delta_{j,i}$ where $\delta_{i,j}$ is 1 if $i = j$ and 0 otherwise. For this special case $C^* \equiv H(X)$, with $P(y_k) = P(x_k)$ for all k. Information is thus transmitted without error when the channel becomes 'typical' with respect to the fixed message distribution $P(X)$.

If $M < L$ matters reduce to this case, but for $L < M$ information must be lost, leading to Rate Distortion limitations.

Thus modifying the channel may be a far more efficient means of ensuring transmission of an important message than encoding that message in a 'natural' language which maximizes the rate of transmission of information on a fixed channel.

We have examined the two limits in which either the distributions of $P(Y)$ or of $P(X)$ are kept fixed. The first provides the usual Shannon Coding Theorem, and the second a tuning theorem variant, that is, a tunable, retina-like, Rate Distortion Manifold, in the sense of Glazebrook and Wallace (2009).

As described above, this result is essentially similar to Shannon's (1959) observation that evaluating the rate distortion function corresponds to finding a channel that is just right for the source and allowed distortion level.

6.8 Some Topological Remarks on Symmetry-Breaking

The fundamental dogma of algebraic topology (e.g., Hatcher 2001) provides instruction regarding the forming of algebraic images of topological spaces. The most basic of these images is the fundamental group, leading to van Kampen's Theorem allowing the construction of the fundamental group of spaces that can be decomposed into simpler spaces whose fundamental group is already known. As Hatcher (2001, p. 40) puts it,

> By systematic use of this theorem one can compute the fundamental groups of a very large number of spaces... [F]or every group G there is a space XG whose fundamental group is isomorphic to G.

Golubitsky and Stewart (2006) argue that network structures and dynamics are imaged by fundamental groupoids, for which there also exists a version of the Seifert–van Kampen theorem (Brown et al. 2011).

Yeung's (2008) results suggest information theory-based 'cognitive' generalizations that may include essential dynamics of cognition and its regulation. Extending this work to the relations between groupoid structures and rather subtle information theory properties may provide one route toward the development of useful statistical models.

References

Albert R, Barabasi A (2002) Statistical mechanics of complex networks. Rev Mod Phys 74:47–97
Brown R (1992) Out of line. R Inst Proc 64:207–243
Brown R, Higgins P, Sivera R (2011) Nonabelian algebraic topology: filtered spaces, crossed complexes, cubical homotopy groupoids. EMS tracts in mathematics, vol 15
Cayron C (2006) Groupoid of orientational variants. Acta Crystallogr Sect A A62:21–40
Cover T, Thomas J (2006) Elements of information theory, 2nd edn. Wiley, New York
Feynman R (2000) Lectures on computation. Westview Press, New York
Glazebrook JF, Wallace R (2009) Rate distortion manifolds as model spaces for cognitive information. Informatica 33:309–346
Golubitsky M, Stewart I (2006) Nonlinear dynamics and networks: the groupoid formalism. Bull Am Math Soc 43:305–364
Hatcher A (2001) Algebraic topology. Cambridge University Press, New York
Horsthemeke W, Lefever R (2006) Noise-induced transitions: theory and applications in physics, chemistry, and biology, vol 15. Springer, New York
Matsumoto Y (1997) An introduction to Morse theory. American Mathematical Society, Providence
Pettini M (2007) Geometry and topology in Hamiltonian dynamics and statistical mechanics. Springer, New York
Shannon C (1959) Coding theorems for a discrete source with a fidelity criterion. Inst Radio Eng Int Conv Rec 7:142–163
Shirkov D, Kovalev V (2001) The Bogoliubov renormalization group and solution symmetry in mathematical physics. Phys Rep 352:219–249
Stewart I (2017) Spontaneous symmetry-breaking in a network model for quadruped locomotion. Int J Bifurc Chaos 14:1730049 (online)
Van den Broeck C, Parrondo J, Toral R (1994) Noise-induced nonequilibrium phase transition. Phys Rev Lett 73:3395–3398
Van den Broeck C, Parrondo J, Toral JR, Kawai R (1997) Nonequilibrium phase transitions induced by multiplicative noise. Phys Rev E 55:4084–4094
Wallace R (2005) Consciousness: A Mathematical Treatment of the Global Neuronal Workspace Model. Springer, New York
Wallace R (2011) On the evolution of homochriality. Comptes Rendus Biol 334:263–268
Wallace R (2016) Subtle noise structures as control signals in high-order biocognition. Phys Lett A 380:726–729
Wallace R (2017) Computational psychiatry: a systems biology approach to the epigenetics of mental disorders. Springer, New York
Wallace R (2020a) How AI founders on adversarial landscapes of fog and friction. J Def Model Simul. https://doi.org/10.1177/1548512920962227
Wallace R (2020b) Signal transduction in cognitive systems: origin and dynamics of the inverted-U/U dose-response relation. J Theor Biol 504. https://doi.org/10.1016/j.jtbi.2020.110377
Wallace R (2020c) Cognitive dynamics on Clausewitz landscapes: the control and directed evolution of organized conflict. Springer, New York

Weinstein A (1996) Groupoids: unifying internal and external symmetry. Not Am Math Assoc 43:744–752

Wilson K (1971) Renormalization group and critical phenomena. I renormalization group and the Kadanoff scaling picture. Phys Rev B 4:317–83

Yeung H (2008) Information theory and network coding. Springer, New York

Printed in the United States
by Baker & Taylor Publisher Services

Printed in the United States
by Baker & Taylor Publisher Services